Tree Whispering: Trust the Path
Notebook and Journal

Jim Conroy, PhD, *The Tree Whisperer®*
Basia Alexander, *The Chief Listener*

and
YOU!

Plant Kingdom Communications™
www.PlantKingdomCommunications.com

First published in 2011 by Plant Kingdom Communications, Publisher.
P.O. Box 90, Morris Plains, NJ, 07950, www.PlantKingdomCommunications.com

The mission of Plant Kingdom Communications is Peace and Balance Among the Beings of Earth.

Copyright ©2012 Jim Conroy, Basia Alexander, and Plant Health Alternatives, LLC.

ISBN-13: 978-0-9834114-1-3

Library of Congress Control Number: 2011932631

Plant Kingdom Communications books may be purchased for educational, business, or sales promotional use. For information, please write: Special Markets Department, Plant Kingdom Communications, LLC, P.O. Box 90, Morris Plains, NJ 07950 USA.

The Plant Kingdom Communications Speaker's Bureau can bring authors to your live event. For more information, please go to *www.PlantKingdomCommunications.com*

Printed in the U.S.A. by Thomson-Shore, Inc., www.ThomsonShore.com
Printed on recycled paper.

Background front cover photo: Shutterstock.com.

Back cover author photos: ©Jane B. Kellner, Kelley/Kellner Associates, Salisbury, CT, kel.kel@sbcglobal.net.

Covers by Basia Alexander and Jim Conroy. Interior design by Basia Alexander.
Copyeditor: Melissa Atkin. Proofreader: Doreen DiGiacomo.

More information about Tree Whispering graduates quoted in this notebook/journal can be found in the Citations of *Tree Whispering: A Nature Lover's Guide to Touching, Healing, and Communicating with Trees, Plants, and All of Nature*, ISBN 978-0-9834114-0-6

The following trademarks are the properties of their respective owners:

Tree Whispering®	*Green Centrics*™	*Leave Leaves*™
The Tree Whisperer®	*Co-Existence Technologies*™	*Botany in Balance*™
Plant Whisper®	*Tree Ambassador*®	*Apple*®
Healing Whispers™	*Plant Ambassador*®	*BodyTalk System*™
Holistic Chores™	*Tree Protectors*™	*HeartMath*®
Strengthen Forests™	*Have You Thanked*™*a Tree or Plant*	
Cooperative BioBalance®	*Today?*	

In gratitude to
our graduates,
our supporters,
the visionaries,
people who protect trees,
those who already talk to plants,
and all the people who have the
courage to do this work.

Who Will Enjoy Using this Notebook/Journal:

❏ Tree people and plant lovers.

❏ People who talk to their plants or trees.

❏ Anyone who has a favorite tree or plant.

❏ Open-minded people. Innovators. Visionaries.

❏ Gardeners, hikers, outdoor enthusiasts.

❏ Spiritually oriented people.

❏ Environmentalists. Sustainability supporters.

❏ Creative spirits. Compassionate, helpful people.

❏ Professionals, farmers, growers, or orchardists.

❏ Folks who already "whisper" with cats, dogs, or other Beings of Nature.

❏ People who feel sad when they see a tree cut down.

When to Use this Notebook/Journal:

❏ When you want to expand your well-being and deepen your appreciation for Nature.

❏ When the trees or plants have a message for you, and you want to capture the experience.

❏ Specifically, when you want to help a particular tree or plant regain its health.

❏ If you enjoy writing, use this notebook/journal when you want to journal about communicating with Nature.

Tree Whispering: Trust the Path
Notebook and Journal

Contents

The most important piece of advice is to trust. Trust that it is our gift and our birthright as humans to be able to communicate with trees and plants. The ability is within all of us.

CATHY, NEW JERSEY, *ACUPUNCTURIST AND HERBALIST*

It seems to me that our biggest stumbling block is self-doubt. We have a great capability to negate our innate intuitive abilities. We have also tuned out [our senses] in order to manage in this fast-paced world. If we were open all the time, we would be overwhelmed. But, I think that when we approach Nature with the intent to ask It about Itself, feelings and words will come forth. And, if we do it enough, then the doubt is eliminated.

MARY MCNERNEY, LINCOLN, MASSACHUSETTS, *ATTORNEY*

I am more aware now of sharing energies with trees. I think the most important thing about feeling a tree's bioenergy field is to believe in the impressions you are getting. I advise people, "Trust that you didn't make it up." I figure, if you get impressions from people when you meet them, why not get an impression from a tree when you meet it?

GEORGETTE HRITZ, SCOTCH PLAINS, NEW JERSEY, *POSTAL WORKER AND HOMEOWNER*

I trust the path in my work. What I don't do is expect to get a predictable answer based solely on my professional training. Professional training teaches that identifiable conditions require certain answers or prescribe certain actions.
Since what I do is beyond conventional diagnosis, I have no preference or agenda when I approach a tree. More than that, I have learned to check my ego at the door. Therefore, I respect a "no."
Hearing "no" does not challenge my personality, and I am not afraid to hear it. I know that trees are not like petulant children. "No" is a piece of good information that leads me to be more creative and insightful because I have to keep asking more carefully worded and deeper questions until I get a better understanding. A "yes" leads me to the next phase of questioning or leads me to clarity.
Sometimes a "yes" or "no" doesn't initially fit my idea of the way I think it should be, but later on in the process, it all becomes clear. I must trust the path.

DR. JIM CONROY, THE TREE WHISPERER

WELCOME!

Dr. Jim Conroy, The Tree Whisperer®, welcomes you: I hope that you are enjoying reading our main book *Tree Whispering: A Nature Lover's Guide to Touching, Healing, and Communicating with Trees, Plants, and All of Nature* shown here.

It is full of do-it-yourself exercises called "Try This." When finishing each exercise in the main book, you are encouraged to write some notes about your experience in this *Tree Whispering: Trust the Path Notebook and Journal*.

We strongly recommend that you use the main book and the notebook/journal together.

Why? The main book provides a frame of reference and deep understanding about the whole topic of communicating with trees, plants, and all of Nature. Without the context and encouragement in that book, you might wonder what these exercises will mean. Or, worse, you might become discouraged.

The exercises are repeated in this companion notebook/journal for your convenience, and with added space to write down your reflections.

Basia Alexander, the Chief Listener, encourages: You may already talk with your plants. If you have had childhood or adulthood connections with trees and plants, this notebook/journal will help you believe in the truth of those experiences. Your perceptions are real! Validate those perceptions and be confident! You can co-create a path to share communication and have a deeper understanding of Nature.

The Exercises Follow This Format

Its name. ──────────→

A reminder to do
only what feels
good and right.

Many steps
with directions.

A reminder to
write your notes in
this notebook/journal.

> **Try This: (the name of the exercise will be here)**
>
> *Read this exercise through first and decide whether you want to do it. Don't do anything that would feel bad for you. If you have any concerns, do not do it.*
>
> **Step 1:** Sit down in a comfortable and private place where you won't be interrupted for 10 to 15 minutes.
>
> **Additional steps follow** with directions that look like this example.
>
> **Last Step:** Jot down some notes in your *Tree Whispering: Trust the Path Notebook and Journal.*

Basia adds: Using this notebook/journal builds trust because it aids in the Tree Whispering process. You do the first exercise, which is very simple. Next, you have an experience during that exercise, then reflect on it by writing. Your experience is the path. Then, you do the next exercise, and your path builds.

The Context of Tree Whispering®

Dr. Jim describes the basics of Tree Whispering: Basia and I are confident that you want to embark upon this journey into the Plant Kingdom because you love trees and plants. We assume you agree that trees and plants are living Beings. You may or may not believe that they have intelligence, but you probably feel that they have a lot they can teach us, if we just listen.

Fundamentally, Tree Whispering is about the health and well-being of trees, plants, and the Beings of Nature. One of the techniques in this notebook/journal will guide you to help your own trees and plants become healthier. Tree Whispering techniques are based on the same principles as holistic, complementary energy-medicine approaches for people. By communicating and helping Green Beings become healthier, your insights and connections to Nature will deepen.

Basia provides perspective: Tree Whispering begins with your personal and profound experience of stepping inside the plant's world. You will do this in the exercise called "Stepping Inside the Plant's World." Our graduates tell us that doing the exercise is a life-changing experience that forever improves and enlightens their relationship with trees and plants. They talk about becoming more respectful with their new partners—the trees. The scope of Tree Whispering then expands into feelings of partnership and the desire to cooperate with all the Beings of Nature as equals. By communicating and helping Green Beings become healthier, your own well-being may also improve since it is a two-way connection loop. We are all connected.

COMMUNICATING WITH THE BEINGS OF NATURE

Basia explains: You may already talk to trees and plants. Many people admit to us that they do talk to their green friends. Some even tell us that they receive messages back. Most fundamentally, Tree Whispering is based on the idea that people and Green Beings can communicate with each other.

You don't have to know anything about botany to do the exercises in this notebook/journal. But, it may help to know that members of the Plant Kingdom have been communicating with each other through chemical scents, tactile exchanges, and bioenergy field impulses since photosynthesizing organisms arose on Earth more than two billion years ago. If they have been communicating with each other, shouldn't we be able to communicate with them? Of course!

It was late in planetary evolution that Beings evolved with specific apparatus for perception. For example, the eye developed at least 540 million years ago. We *Homo sapiens* have evolved five physical avenues of perception—seeing, hearing, touching, smelling, and tasting—and also sensitivities to bioenergy field impulses.

Our most sensitive and powerful organ of perception is our heart. The human heart is like another brain: It is made of neural tissue. It also broadcasts and receives like a radio transmitter and receiver. So, people can communicate with Beings of Nature—not through spoken language but through sensory sensitivity, emotional engagement, intuition, and the power of the loving heart. Several of the exercises take you step-by-step through the process of expanding your perceptions and developing intuitive skills.

© Basia Alexander

Dr. Jim tells how he communicates with Green Beings: When I touch a plant or put both of my hands on the trunk of a tree, I feel a surge of energy—like an electrical current—moving through it. My senses and intuition receive its messages as feelings, aromas, or words. Every person has a different specific experience, but all who want to communicate with the Beings of Nature can do so by becoming better receivers. We may call this "Tree Whispering," but it is really "Nature Listening."

Basia offers advice: To become better receivers of communications from any Being of Nature, people need to rise above culturally defined limitations. Many people have been told that humans are not capable of sensory or perceptive expansion. I say, "Why argue in favor of a limitation?" You are far more perceptive and sensitive than you usually believe. Dr. Jim and I strongly encourage people to accept and validate their youthful playtime adventures and adult experiences of connection with the Life Force, bioenergy, Spirit, and wisdom of trees and plants.

Dr. Jim talks about healing trees and plants: I approach trees and plants with the intention of healing them when they are weak or sick. For me, the communication involves asking them a lot of questions and working in an equal partnership with the Green Beings toward a goal. The goal is the trees' recovery and reestablishment of their inner health.

With my hands on a tree, I come from the tree's point of view by stepping inside of its world. Over time, I have come to trust the communication that occurs between the tree and my left brain knowledge, my right brain intuition, and my heart's bioenergy overlap with the tree's bioenergy field. I imagine that it looks like the diagram to the left.

Chapter 4 of the main book gives more detail about human and tree bioenergy fields.

To heal the Green Being, I first ask permission to interact with the Green Being because it is only respectful to ask another living Being before engaging with it. Then I start asking specific and detailed questions from my Green Centrics™ professional healing system. I find out how the tree or plant is sick and how I can heal its compromised internal functionality.

I ask many questions, and the Green Being answers me—not so much in words but in feelings, images, sounds, fragrances, impulses, and other directions telling me how I can apply my healing efforts. With each question, I take a step closer to the healing goal.

Later in this notebook/journal, we will give you the steps of the first Healing Whisper™ healing technique. We feel we are giving this as a gift to you. It is designed to be as simple and easy as possible. It's a healing method you can use effectively on your own trees and plants without going through my advanced and detailed questioning process.

Still, you will be asking your own questions of your Green Being. It's an opportunity to customize Tree Whispering to your situation. Asking questions is one of the key aspects of trusting the path. By asking your questions, you are better able to understand what is going on inside the plant, to communicate with it, and even to help heal it.

© Basia Alexander

THE PATH THROUGH THE WOODS

Basia describes: When you take a quiet walk in the woods, how do you feel? Even now, you can imagine walking and being immersed in the pleasures of the woods: You are surrounded by strong trees, whispering leaves, beautiful colors, musky scents, and patches of warm sunshine.

Don't you feel good—body and soul—when you are connected with Nature? Perhaps on a quiet walk in the woods, you feel or believe that you might leave your world and step into the trees' world.

And so, you can! The main book and this notebook/journal will guide you, but you must eventually take steps into the trees' world yourself.

The deep quiet in the woods—perhaps only punctuated by birdsong or rustling leaves—invites your thoughts and your spirit into a deeply quiet place of inner reflection and contemplation. On your quiet walk in the woods, don't you usually follow a path? Your feet do the task of taking a series of steps on the path in the woods while you step deeper and deeper inside of yourself.

But sometimes, there is no path. You have to make your own path.

Walking becomes a process of discovery.

So, please allow me this metaphor: The Tree Whispering exercises are a series of steps you take not only within yourself but also in partnership and cooperation with your friends—the trees, plants, and other Beings of Nature. In this notebook/journal, *we* are providing step-by-step procedures. *You* are engaging in your own process of discovery and trusting the path you forge with trees, plants, and Beings of Nature.

INSIGHTS INTO TRUSTING

Dr. Jim explains: Many people are comfortable when they feel that they are in control of a situation and when they think that they know what to do. In life, that can be a good strategy for success.

In Tree Whispering, the pressure is off.

You do not have to be in control. You do not have to know what to do. You don't have to set expectations for yourself. You can release any need you may have to be right or to look good.

The exercises you'll try in this notebook/journal are all about allowing *experiences* to happen, receiving feelings or sensory input, taking the tiny and step-by-step risks of experimentation, having a good time, and feeling balanced within yourself.

Basia offers some insights: The word "trust" usually carries a lot of emotional impact. So, I often turn to the dictionary and thesaurus to see the shades of meaning that a word can have. "Belief that something is true" and "placing confidence in something" are generally what Dr. Jim and I are advising when we suggest that you "trust the path" as you are doing the exercises in this notebook/journal.

All too often, we find that people invalidate themselves by thinking "I can't do this." Or, people invalidate a process by deciding that "it won't work for me." Please, don't set yourself up to fail. We encourage you to believe that communicating with trees, plants, and other Beings of Nature is possible—and is possible *for you.*

Dr Jim explains how to trust the path: The path is doing the exercises and having experiences. As you experience the path, you begin to trust the path. Experience is the best teacher. If the experience rings true for you and feels good, then trust it. Don't invalidate it. Then, move on to the next exercise. Please see the diagram on this page.

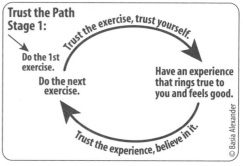

Dr. Jim continues: It is like learning to ride a bicycle. Until you experience balance on the bike, you don't really understand it. You peddle then try to keep the front tire straight. At first, you are wobbly; you may even fall. Somebody else may be holding the bike for you. You keep trying—you are determined—peddling, and keeping the front tire straight. All of a sudden, you go 25 feet easily all by yourself. The experience of balance is growing. Possibilities open up to go one hundred feet, or even to get to the end of the block. The path opens up, your experience of balance is growing.

In order to communicate with trees, plants, and Nature, I think it starts the same way. At first, the exercises may feel different. You may feel unsure of yourself. But, you continue to do the exercises. All of a sudden, you begin to feel the experience of communication opening up. It rings true for you. Then, you do more exercises and your experience grows. Your experience deepens and the path opens up for you because you see more possibilities.

In this case, the trees, plants, and Beings of Nature are doing something like "holding up the bicycle" for you. They want to communicate with you. All you have to do is "trust the path" that they are opening up for you. Your experience grows and deepens. As you continue to do the exercises, and your experiences grow, you may even get the sense that the trees, plants, and Beings of Nature are lighting the path for you. The diagram on the facing page details this stage of development.

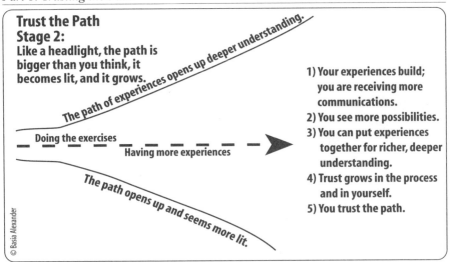

Trust the Path
Stage 2:
Like a headlight, the path is bigger than you think, it becomes lit, and it grows.

The path of experiences opens up deeper understanding.

Doing the exercises

Having more experiences

The path opens up and seems more lit.

1) Your experiences build; you are receiving more communications.
2) You see more possibilities.
3) You can put experiences together for richer, deeper understanding.
4) Trust grows in the process and in yourself.
5) You trust the path.

© Basia Alexander

Basia concludes: Each additional exercise draws you forward into more of the path—which means trusting your experience and trusting yourself. You will gain a deeper understanding of your communication with the Beings of Nature as your experiences accumulate—which makes trusting the path easier.

We suggest that you set aside some private and uninterrupted time to do each exercise. Workshop graduates report that they see trees with new eyes and feel appreciation for plants with open hearts! These short and easy processes are wonderful jumping-off points for reconnecting with the sacredness of Life and having a dialog with any of Nature's Beings including trees, plants, bees, other insects, and even microorganisms.

Then, you'll probably realize that you are finding out what is really important to you and in life. These exercises have a way of helping you reprioritize life circumstances and reestablish well-being.

THE PARTNERSHIP WITH BEINGS OF NATURE

Basia explains: For over 500 years, Western culture has been operating with the mindset that people could control Nature's forces as if they were mechanistic parts, should dominate Nature for humanity's benefit, and that science made people superior. How has that been turning out for us? for the Beings of Nature? and for planetary environments? Not entirely well, I think.

Since we are all connected on this small planet, those assumptions must change. Our ideas and practices need to become focused on partnership with trees and plants, cooperation, and a sense of equality with the Beings of Nature. Our practices need to shift toward co-creative and respectful approaches of cooperation, partnership, and equality.

Dr. Jim explains: Some people don't realize that Nature is out of balance. Other people have accepted that Nature is out of balance and often feel resigned and hopeless about it. People see Mother Nature's response to environmental imbalances as new insects or organisms arising and as weather extremes. Left to Herself, Mother Nature would bring Earth back into dynamic balance given geological time. In doing so, however, humanity might be optional for Her to keep around.

But, I am here to say that there is hope; there is cause for celebration! "Out of balance" is only a model that Western culture has made dominant. But there is another model: I call it "Nature in Balance." We are not helpless and can feel empowered. Each and all of us can—in a relatively short timeframe—turn around the messes humanity has made. Each and all of us can shift our consciousness and change our practices. Using this notebook/journal is a good beginning!

Balance within Nature—dynamic balance—is all about relationships, feedback loops, and network patterns. There are three parts to reestablishing "Nature in Balance."

Dr. Jim describes the three parts: What comes first? Communicating and cooperating with Green Beings comes first. Healing trees and plants from the inside-out and coming from their point of view also comes first, so that trees and plants are healthy. We depend on their good health and support for our very lives.

Secondly, the interrelationships between plants and organisms such as bees and all other insects, diseases, fungi, and the forces of Nature (earth, air, fire, and water) need to be brought into harmony. This is possible through Cooperative BioBalance®.

The Beings of Nature already had balance, harmony, and collaboration. We–Human Beings–have messed it up. Which brings me to the third part needed to reestablish "Nature in Balance": it is about each of us transforming our relationships with the Beings of Nature as well as changing our attitudes and practices to cooperation, partnership, and equality. Transforming and changing our ways by using Cooperative BioBalance is good for the planet we live on and share with all Beings.

The new field of study called Cooperative BioBalance rests on these principles.
  • *Live and let live.*
  • *We are all connected.*
  • *People can communicate with trees and plants and other Beings of Nature.*
  • *Ask first, come from the plant's point of view—don't assume you know.*
  • *Trees and all plants can live in dynamic balance with diseases, insects, and other organisms.*

Dr. Jim concludes: The Institute for Cooperative BioBalance will found and ground a new field of study called Cooperative BioBalance. It is the study and co-creative practice of dynamically balanced relationships among all living Beings. This means partnering with the Beings of Nature equally and for mutual good. It is about each and all of us reconnecting with Nature, coming from the plant's point of view, and bringing wildly fluctuating ecosystems back into dynamic balance.

Your study of Cooperative BioBalance begins here and now through your partnership with Green Beings in this notebook/journal. It is in your hands. Balance in Nature is also—ultimately—in your hands.

OTHER ADVICE FOR YOUR JOURNEY ON THE PATH

Basia talks about boundaries: Once you get to know a tree or plant, you can receive information about its health or get messages from it. You should be receiving positive information and pleasant sensations, not pain or discomfort of any kind. If you feel pain inside your body, stop the process immediately. Develop strong, healthy personal and psychological boundaries before you communicate with a tree or plant, or work to heal a sick tree or plant.

Basia adds: We designed this notebook/journal to be carryable. So, please carry it with you so you'll be ready anytime to get trees' and plants' messages. It is a convenient size to take along on a quiet walk in the woods. Or, make sure it is handy when you are helping trees and plants become healthier with a Holistic Chore™ or Healing Whisper™.

Graduates from our classes have said inspiring things about their relationships with trees and plants. You'll find their words on various pages. Their words offer insight into the sections where they are found in this notebook/journal. More information about the graduates is located in the citations at the back of our main book *Tree Whispering: A Nature Lover's Guide to Touching, Healing, and Communicating with Trees, Plants, and All of Nature.*

Dr. Jim reminds: The trees are waiting for you to communicate with them. When you have an open mind and an open heart, you might receive a message from any Being of Nature. When one has a message for you, write it down; write immediately or you might forget the message. Whenever you appreciate Nature, it's easy to write a few notes about how you feel or what you think.

Basia reflects: My friend, Doreen, who practices a form of energy medicine told me, "I used to worry that I might not 'get it right' with a client. But now, I realize that I am not doing the healing. A higher energy—combined with the intention that my client and I hold—is doing the healing. It is reassuring to know that I just have to step out of the way and let it happen." So, doing these exercises and Tree Whispering can be easy. You don't need memory. You are tapping into a place where you need little or no instruction. You just need yourself. Confidence will grow in both your skills and the exercises. And, please, have a good time.

Don't be afraid of the blank page. Here's a little secret: Sit quietly with your pen poised on the paper. Then, begin writing some pleasant words just to get the pen moving. Once the pen is moving, words can become sentences. Soon, you'll fill the page.

KEEP EXPANDING

Dr. Jim adds: Even if you haven't read the main book, this *Tree Whispering: Trust the Path Notebook and Journal* can be an amazing journey, so don't stop now. We will give explanations along the way to guide you on the path.

You can find many of this notebook/journal's exercises available for download as MP3s at *www.TreeWhispering.com*.

If you are inspired to creativity, aroused to activism, or interested in any additional activities, please refer to the section called "Participate" at the back of this notebook/journal.

If you want to get to know more about Basia and me, more information is at the back of this notebook/journal, too.

USE THIS NOTEBOOK/JOURNAL TO
- Allow experiences to happen
- Receive feelings or sensory input
- Take the tiny and step-by-step risks of experimentation
- Have a good time
- Feel balanced within yourself
- Write about your reflections

Being perceptive? Well, we have to get over our constant second guessing. I say to you, "Trust your first impression always."
    LIZ WASSELL, NEW PALTZ, NEW YORK, COPY EDITOR, REIKI PRACTITIONER, ANIMAL AND NATURE COMMUNICATOR

My advice for people about being better receivers of communications from trees and plants is simple: Keep an open mind. Have no distractions. Be present in the moment. Your "now" presence is required to be with Nature. Be mindful of where your focus is directed.
    SYLVIA D'ANDREA, NEW JERSEY, GRAPHIC DESIGNER AND HOMEOWNER

Trust that you will receive what you need. Set the intention that your intuition will expand, let your imagination take over, and have fun with it. Some like to use journaling to bring information through and others use meditation. By practicing journaling or meditation , you will more quickly connect to the Divine, all-knowing, aspect of yourself.
    LINDA S. LUDWIG, USA, DIVINE HEALER AND BUSINESS OWNER

TRY THIS: FEEL GOOD VISITING A TREE OR PLANT

Read through the steps. Decide whether you feel comfortable doing this exercise. If you do not feel comfortable or have any concern whatsoever, do not do it.

Step 1: If it is practical, go to your favorite tree or plant. Or, go to any tree or plant you like or think is beautiful.

Step 2: Allow yourself the pleasure of feeling good in its presence. Don't try to think, analyze, or understand anything. Just feel good. Enjoy yourself.

Step 3: Slow and deliberate breathing can help. Stay with it for five to ten minutes.

Step 4: Jot down some notes to yourself. Focus on your impressions and on how good you feel. If necessary, sit quietly with your pen poised on the paper. Try writing some pleasant words just to get the pen moving. Once you are writing the first words, more words will come.

TRY THIS AGAIN: FEEL GOOD VISITING A TREE OR PLANT

Read through the steps. Decide whether you feel comfortable doing this exercise. If you do not feel comfortable or have any concern whatsoever, do not do it.

Step 1: If it is practical, go to your favorite tree or plant. Or, go to any tree or plant you like or think is beautiful.

Step 2: Allow yourself the pleasure of feeling good in its presence. Don't try to think, analyze, or understand anything. Just feel good. Enjoy yourself.

Step 3: Slow and deliberate breathing can help. Stay with it for five to ten minutes.

Step 4: Jot down some notes to yourself. Focus on your impressions and on how good you feel. Begin writing some pleasant words just to get the pen moving. Once the pen is moving, words can become sentences. Soon, you'll fill the page.

_____

_____

_____

_____

_____

_____

_____

_____

_____

_____

_____

_____

_____

_____

_____

_____

The more I practice awareness, the more my own feelings open up. I often ask questions of the trees, and their answers come in song. I suggest that people go out, sit under a tree, breathe and relax, listen and feel. Most importantly, be patient. Let thoughts come and go; be quiet. Try touching the tree—touching helps in the beginning. Pick one tree and visit it often. It will feel more familiar each time. Simply have fun!

LORI MYRICK, EAST WINDSOR, CONNECTICUT, ENERGY THERAPY PRACTITIONER

TRY THIS: A QUIET WALK IN THE WOODS

Read through the steps. Decide whether you feel comfortable doing this exercise. If you do not feel comfortable or have any concern whatsoever, do not do it.

Step 1: Take your *Tree Whispering: Trust the Path Notebook and Journal* with you to your favorite park, woods, forest, or path through trees.

Step 2: Begin your walk in the quiet.

Step 3: In your heart or speaking in a whisper, express gratitude to the trees as you move among them. Notice how you feel.

Step 4: Stop for a moment. Say to the trees: **"I open my heart to you."**

Step 5: Resume walking. Be receptive to feelings of their gratitude coming to you.

Step 6: As you walk or afterward, jot down notes about how you feel, what you think, and what it was like for you to take this walk. Writing brief phrases is okay.

I suggest that you bring a notebook and pens or pencils when you connect with the energy of trees and plants. You should record the messages you get immediately. If you are receiving words, write them. If you are inspired to draw, then draw. Just record the experience as you are getting it without engaging the rest of your mind in that busy-talk that can squash the communication.

LIZ WASSELL, NEW PALTZ, NEW YORK, COPY EDITOR, REIKI PRACTITIONER, ANIMAL AND NATURE COMMUNICATOR

TRY THIS AGAIN: A QUIET WALK IN THE WOODS

Read through the steps. Decide whether you feel comfortable doing this exercise. If you do not feel comfortable or have any concern whatsoever, do not do it.

Step 1: Take your *Tree Whispering: Trust the Path Notebook and Journal* with you to your favorite park, woods, forest, or path through trees.

Step 2: Begin your walk in the quiet.

Step 3: In your heart or speaking in a whisper, express gratitude to the trees as you move among them. Notice how you feel.

Step 4: Stop for a moment. Say to the trees: **"I open my heart to you."**

Step 5: Resume walking. Be receptive to feelings of their gratitude coming to you.

Step 6: As you walk or afterward, jot down notes about how you feel, what you think, and what it was like for you to take this walk. Let the words flow from your heart.

_____

_____

_____

_____

_____

_____

_____

_____

_____

_____

_____

_____

_____

_____

_____

continues...

There are life lessons and wisdom all around, encircling me as I take my quiet walk in the woods.

Aged trees that stand tall, erect, and branching widely over tender saplings, teach me the importance of elders who guard, protect, and shade their young from harm. Broken limbs that fall onto the branches of neighboring trees can remain there for years, just as people support their weakened and failing loved ones for as long as it is necessary. I learn patience from passing the same familiar trees year after year. Just as children pass through phases of growth and maturity, the trees bud every year with new life, flower with enthusiasm, transform from deep greens to golden autumn shades of russet, and endure winter's cold and darkness with noble tranquility.

Just looking at the differences between trees standing side by side illustrates how every living Being is an individual with its own characteristics. The varied texture of bark is like my unique fingerprint, none exactly like another. As my stance differs from that of others, so too does the stature of trees show marked contrasts. An oak can stand upright like a soldier on guard duty, the white birch can bend low like an elderly grandmother, and the evergreen can bow with a sweep of its boughs like a princess swinging her gown.

I find that the more I pay attention to trees, the more I appreciate them and respect them for what they add to my life, and I am grateful for their presence.

ANN ST. GERMAINE, EATONTOWN, NEW JERSEY, EDUCATOR, WRITER AND PHOTOGRAPHER

TRY THIS: LOCATE YOURSELF ALONG THIS CONTINUUM

Read through the steps. Decide whether you feel comfortable doing this exercise. If you do not feel comfortable or have any concern whatsoever, do not do it.

Step 1: Identify where your ideas, beliefs, attitudes, and concepts are along this continuum:

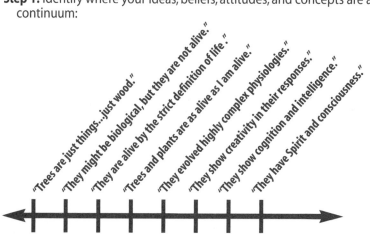

Step 2: Make a note below about where your ideas fall along this continuum.

_____

Step 3: If the statements don't resonate with you, write down your own beliefs or attitudes about the aliveness of trees and plants.

_____

Step 4: Ask yourself: **"Am I willing to shift my attitudes or beliefs?"** Jot down your answer.

_____

Step 5: Ask yourself: **"What would happen in my life if my beliefs changed?"** Think about this for a moment and write your response.

_____

_____

Step 6: Ask yourself: **"Would I allow my beliefs to shift further toward the right side of this continuum?"** Jot down your first response.

_____

_____

Step 7: Write any additional notes or reflections.

_____

_____

_____

_____ *continues...*

My advice is to recognize the consciousness in Nature. The best information does not come through thoughts and mind; it comes through feelings. The greatest percentage of my work is done through feelings, and then I interpret those into thoughts that my mind can use.
JEFF DAWSON, NAPA, CALIFORNIA, HORTICULTURALIST

I practice and teach a healing system that is fundamentally based upon communicating with the innate intelligence within the human body. Developing one's perceptual skills and being a good receiver of communication in complementary energy healing means acknowledging that all life has consciousness.
MELANIE BUZEK, CORNVILLE, ARIZONA, PHYSICAL THERAPIST, ENERGY-MEDICINE PRACTITIONER AND EDUCATOR

TRY THIS: AWARENESS OF YOUR HEART FIELD

Read through the steps. Decide whether you feel comfortable doing this exercise. If you do not feel comfortable or have any concern whatsoever, do not do it.

Please see a detailed explanation of human heart fields in the Appendix of our main book Tree Whispering: A Nature Lover's Guide to Touching, Healing, and Communicating with Trees, Plants, and All of Nature.

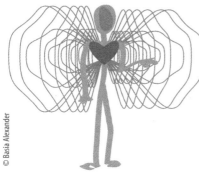

© Basia Alexander

Step 1: Sit down in a comfortable and private place where you won't be interrupted for five to ten minutes. This exercise is best done with eyes closed, if you are comfortable closing your eyes.

Step 2: You have a donut-shaped bioenergy field around you. Pause for a moment and allow yourself to experience your own bioenergy field radiating from your heart.

Step 3: Imagine that you can see your bioenergy field's shape or feel its warmth.

Step 4: If you are within about five feet of another person, please take a moment to realize that both of your bioenergy fields are overlapping and interacting.

Step 5: Think about what having a bioenergy field generated by your heart means to you.

Step 6: Think about the fact that bioenergy fields can overlap. What does that mean to you?

Step 7: Jot down a few notes about your experience. If necessary, sit quietly with your pen poised on the paper. Try writing some pleasant words just to get the pen moving. Once you are writing the first words, more words will follow.

_____

_____

_____

_____

_____

_____

_____

_____

_____*continues...*

Developing the heartspace is the most important aspect for me. The heart is a perceptive tool. As I learned from Dr. Jim and Basia, the HeartMath Institute has done leading-edge research into the ways the human heart works as a receiver of Nature's communications. I use meditation for developing my conscious connection into the heartspace.

JEFF DAWSON, NAPA, CALIFORNIA, HORTICULTURALIST

When I was in graduate school, there was little credence given to the mind-body connection. But, things have changed. Now we know that giving up focus on self and opening the heart in communication with other Beings results in benefits for a person that can be measured as reduced inflammation, less anxiety, and less distress. So it works both ways: by helping trees I feel like they are helping me. *ANONYMOUS, PHD PSYCHOLOGIST*

TRY THIS: INVITE SENSORY SENSITIVITY

Read through the steps. Decide whether you feel comfortable doing this exercise. If you do not feel comfortable or have any concern whatsoever, do not do it.

Audios are available to download at
www.TreeWhispering.com

Note 1: You may repeat this exercise regularly, as long as you feel good and enjoy it.

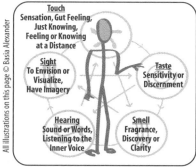

Right-Brain: Processes input. Intuitive and Holistic.

Left-Brain: Processes input. Logical and Hierarchical.

HeartBrain: Perceives. Transmits and Receives. Produces Bioenergy Field.

Note 2: Making notes each time you do this exercise will give you a sense of your progress.

Note 3: You may or may not feel an immediate enhancement of the chosen sense. The exercise is designed to provide the foundation to the bodymind in which the enhancement can occur. The exercise will not by itself cause the sense to be enhanced.

Step 1: Sit down in a comfortable and private place where you won't be interrupted for five to ten minutes. This exercise is best done with eyes closed.

Touch
Sensation, Gut Feeling, Just Knowing, Feeling or Knowing at a Distance

Sight
To Envision or Visualize, Have Imagery

Taste
Sensitivity or Discernment

Hearing
Sound or Words, Listening to the Inner Voice

Smell
Fragrance, Discovery or Clarity

All illustrations on this page © Basia Alexander

Step 2: Choose a sense that you want to enhance. (See illustration to the left.) Jot it down here: _____

_____

_____

Step 3: Focus on the sense and take a comfortable, deep breath.

Step 4: Gently tap with the fingertips of **both hands** on the heart (center of the chest).

Continue to take 3 to 4 comfortable breaths while you **tap on the heart** and **focus on the sense.**

Step 5: Gently tap with the fingertips of **both hands**, each on opposite areas of the head. You may tap anywhere: front, top, back, or sides. Do this while taking 3 to 4 comfortable breaths. Focus on the sense.

Step 6: Repeat Steps 4 and 5 about five times while focusing your attention on the sense.

Step 7: Open your eyes, breathe gently, and sit quietly for a moment.

Step 8: Jot down any comments or reflections on the next page.

continues...

I use a little secret to expand my trust in my heart. It is based on the tapping exercises Basia showed us in the workshop. I tap on the center of my chest while I give my heart energy-center permission to 'be' and permission to have its own say. Then, I tap between my eyebrows and tell my heart's energy to be uppermost while it links with my headbrain. I do this tapping back and forth from heartbrain to headbrain many times, asking them to work in conjunction. It helps me settle down. I am grateful. GERRY VERRILLO, GUILFORD, CONNECTICUT, ARBORIST

TRY THIS: ENHANCE EMOTIONAL ENGAGEMENT

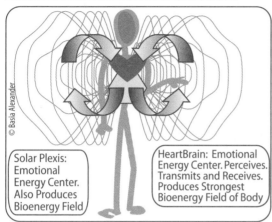

© Basia Alexander

Solar Plexis: Emotional Energy Center. Also Produces Bioenergy Field

HeartBrain: Emotional Energy Center. Perceives. Transmits and Receives. Produces Strongest Bioenergy Field of Body

Read through the steps below. Decide if you feel comfortable doing this exercise. If you do not feel comfortable or have any concern whatsoever, do not do it.

Note 1: If possible, use your nose for breathing IN and your mouth for breathing OUT during this exercise. Best done with eyes closed.

Note 2: You may repeat this exercise regularly, as long as you feel good and enjoy it.

Note 3: Making notes each time you do this exercise will give you a sense of your progress.

Note 4: Engage this pair of emotions: courage and gratitude. You may return to this exercise any time with any pair of positive and life-affirming emotions.

Step 1: Sit down in a pleasant and private place where you won't be interrupted for about 10 minutes.

Step 2: Focus your attention on the feeling or idea of courage.

❑ *Remember a time when you felt courageous or brave. Or, recall a scene from a movie in which you felt that a character was doing a courageous act.*

❑ *Take a moment to feel courageous or brave.*

Step 3: Focus your attention on the feeling or idea of gratitude.

❑ *If there is something you are grateful for, think of that and allow the experience of gratitude to come to you.*

❑ *Take a moment to feel grateful.*

Step 4: Imagine a stream of air looping in through your heart area and out through your belly area.

Step 5: While imagining the looping stream of air, gently and easily breathe in and out for 4 or 5 breaths.

Step 6: Imagine that on that stream of air, you are breathing in **courage** through your heart and breathing out **gratitude** through your belly area.

❑ *Allow yourself to gently experience courage flowing in, then gratitude flowing out.*

❑ *Or allow yourself to simply relax and enjoy breathing in and out slowly and gently. Do this for about a minute.*

❑ *If you are comfortable and enjoy this experience, you may continue for several more minutes.*

Step 7: When you feel ready, open your eyes and sit quietly for a moment.

Step 8: Jot down any comments or reflections on the next page.

continues...

When I am out in wild Nature, I practice gratitude for living Beings and all of Nature.
    I would advise people first to see the possibility of communication, then to have confidence
while being open-minded and open-hearted. Follow up with gratitude. People need to know that
this Nature communication can be done simply, easily, and that it enhances daily living.
        CAROL OHMART-BEHAN, ENDICOTT, NEW YORK, AUTHOR AND GUIDE OF SPIRITUAL JOURNEYS

TRY THIS:
"STEP INSIDE THE PLANT'S WORLD" GUIDED VISUALIZATION EXPERIENCE

Dr. Jim says: This is the key exercise for getting in touch with the energy and Life Force of trees and plants. It may seem like it has a lot of steps, but, please, enjoy yourself while you go through it. You are stepping into a new world and can come back with feelings of inspiration and a great story to tell! This exercise is the most effective way to begin communicating with Green Beings.

Read through all the steps. Decide whether you feel comfortable doing this exercise. If you do not feel comfortable or have any concern whatsoever, do not do it.

Audios are available for download at www.TreeWhispering.com.

Preparation: Make sure you are in a pleasant and private place where you won't be interrupted for about 15 minutes. There will be times during the exercise that you may want to close your eyes. If you are comfortable doing that, please do so.

• Make sure you have a plant near you that you can touch. You may also do this exercise standing or sitting outside privately with a tree.

• Have your notebook/journal handy to write down some notes about your experiences. It's best to make notations immediately in case you don't remember later.

• Questions will be asked during the experience. Those questions are meant to be answered privately, inside yourself. You may jot notes for answers.

• Please remember this illustration as you do the exercise.

© Basia Alexander

Step 1: Begin.

❏ *Get comfortable.*

❏ *Breathe deeply and gently.*

❏ *Turn your attention to the plant.*
Never mind its name. Forget anything that you may know about it.

Step 2: Since it is a living Being, always ask permission to make contact.

❏ *In your Heart, say to the plant or tree,*
"I would like to spend some time with you and get to know you. Is that okay?"

❏ *You will probably feel a sense of calm. That means you have permission from the plant. It is unlikely, but if you feel disquiet, you may move to another plant, or stop.*

Step 3: Focus your sight on the stem, trunk, leaves, or flowers. Notice even the tiniest details.

❏ *See a multitude of shapes.*

❏ *Notice how the leaves are arranged on the stem.*

❏ *Notice how leaves or stems are oriented in space.*

❏ *Distinguish shades of color.*

❏ *See how light or shadow plays around them.*   *continues...*

Step 4: Now, very gently, touch the plant or tree.

❏ *Feel the textures. Notice the shapes.*

❏ *Perceive curves and turns.*

❏ *Sense the temperature.*

Step 5: Use a soft focus and perceive the whole plant or tree.

Step 6: Ask yourself: **"How do I feel?"** and **"What do I know?"**
Quietly answer those questions for yourself or jot down a few notes on page 28.

Step 7: Focus for a moment on your own body. You may close your eyes.

❏ *Around your body, notice your heart's bioenergy field.*

❏ *Notice the bioenergy field's current size and shape.*

❏ *Imagine that your heart's biofield is like the sun—shimmering with light or radiating with energy.*

❏ *Imagine that the field is increasing in size and intensity.*

Step 8: Notice or imagine the bioenergy field of the plant or tree. You may close your eyes to do so.

❏ *Imagine that the plant's or tree's bioenergy field is also expanding. Feel it.*

❏ *Your heart's field now overlaps with the energy field of the plant or tree.*

❏ *In that overlapping area, information is shared and exchanged, sensory experience is stronger, and emotional perception is heightened.*

Step 9: With your eyes closed, for about a minute, allow yourself to be aware of any new information, experience, or perception in the bioenergy overlapping state (as the illustration on the facing page suggests).

Step 10: Step into the plant's or tree's world.

❏ *Engage your imagination and close your eyes.*

❏ *Feel as if you shrink or expand to fit the plant's or tree's size.*

❏ *Imagine that there is a door on the stem or trunk. You open it. A bright, white light shines on you, and you step inside the stem or trunk.*

❏ *There are thousands of cells all around you. Imagine reaching out with your hands and touching some cell walls.*

❏ *As you look around, you see little streams going up and down all around you. These are circulating plant fluids.*

❏ *Imagine that a tiny boat comes along. Hop in and begin moving upwards with the flow of the fluids. Continue moving upward.*

Step 11: Ride into a leaf.

❏ *You approach a leaf. The boat docks. You float forward, following a vein into the leaf.*

❏ *Sense light coming through the upper layers of cells.*

❏ *There is a lot of activity: A bubble of carbon dioxide is captured. A bubble of oxygen is released.*

❏ *Sense the heat from foods being produced.*

❏ *You are inside the leaf.*

Step 12: Ask yourself: **What do I sense?   What do I notice?   What do I realize?"** and **"What is important to the plant?"**
Quietly answer those questions for yourself or jot down a few notes on page 28.

The exercise continues on the next page...

Step 13: Ride the sugar molecule.

❑ *Inside the leaf, photosynthesis is producing sugar molecules, which are food for the plant. Imagine that you shrink down so small you can jump on a sugar molecule.*

❑ *Ride it into the stream. Travel out of the leaf and down the stem or trunk.*

Step 14: Move inside the roots.

❑ *Imagine that the stream flows beneath the soil and you gently submerge into the root zone. Imagine that–ahead–the little stream is splitting and narrowing.*

❑ *Your molecule stops on the side of the little stream as food.*

❑ *You step off the sugar molecule. You are inside the roots.*

Step 15: Ask yourself: **"What do I sense?   What do I notice?   What do I realize?"** and **"What is important to the plant?"**
Quietly answer those questions for yourself or jot down a few notes on page 28.

Step 16: Pick up the sponge.

❑ *Imagine turning around. Another stream is behind you. This one is flowing upwards.*

❑ *Now, imagine that you have a sponge. Step toward the stream, put in the sponge and sop up some water and nutrients. Another little boat comes along. Bring the sponge with you and get in.*

❑ *Feel a surge forward as the root's pumping action pushes you upward. You emerge and are traveling in the brightness, up the stem or trunk again.*

❑ *This stream flows toward a special part called the "growing point." Once there, you deposit the water and nutrient-soaked sponge in a growing cell.*

❑ *Imagine that the cell begins to divide and grow.*

Step 17: Fill the whole plant or tree.

❑ *Now, feel yourself expand and grow until you fill the whole plant or tree.*

❑ *Sense all the activity—thousands of things happening and interacting. All these interactions and feedback loops are its Growth Energy.*

❑ *Almost like a pulse or heartbeat, sense the strength of that Growth Energy.*

Step 18: Come from the plant's or tree's point of view.

❑ *Feel the movement of the Growth Energy around the plant or tree.*

❑ *Perceive where it is surging. Find out where it might be weak.*

Step 19: Ask yourself: **What do I sense?   What do I notice?   What do I realize?"** and **"What is important to the plant?"**
Quietly answer those questions for yourself or jot down a few notes on page 28.

Step 20: Enjoy a few moments with the tree or plant.

Step 21: BEING the tree or plant and the Flow of Collaboration

❑ *Realize that you have made a new friend and can do something like "see" through its "eyes" or "feel" through its life. Feel the two-way flow of connection between you and the tree or plant.*

❑ *Notice that you are coming from the plant's or tree's point of view and not coming so much from your own point of view.*

❑ *Acknowledge that you and the plant or tree have created a conscious partnership. Feel the collaboration between the two of you like a two-way street.*

Enjoy this experience for a few moments. Jot notes on the next page.

continues...

Step 22: One at a time—slowly and gently—ask these questions to the plant or tree and immediately write down any feelings or impressions you receive. Your unique perception may be to hear it, see it, or feel it.

❏ Ask: *"Do you have an instruction for me?"* Jot any notes below.

❏ Ask: *"Do you have a message for me?"* Jot any notes below.

❏ Ask: *"Do you have a lesson for me?"* Jot any notes below.

Step 23: Exchange gifts with the plant or tree.

❏ Breathe in any gift(s) from the plant.

❏ Breathe out and say *"Thank You"* to the plant for letting you into its world.

Step 24: Leave the plant or tree.

❏ If you are still touching the plant, slowly release it.

❏ Return to your own body and to your own point of view. When you are ready, wiggle your toes. Take in a long, deep breath and stretch your back and shoulders.

Step 25: If you have not already written anything, please write your answers to these questions below. If necessary, sit quietly with your pen poised on the paper. Try writing some pleasant words just to get the pen moving.

❏ What was important to the plant?

❏ What were any messages, instructions, or lessons from the plant?

❏ What gift did you receive?

Step 26: Share your experience with other people.

❏ Think of a friend or family member with whom you may share your personal experience. Imagine sharing this experience with them.

❏ If you feel safe, actually tell them about it. If you don't feel safe or think it would be a bad idea, don't do it.

-----

continues...

continues...

*In connecting with trees' root systems, I understand much more fully how trees stand for us,
marking time, generating oxygen while observing our actions, and holding energy in their
roots and in the soil from which they are fed. They are helping us, and we are partners.
For their gifts, they respond profoundly to simple acknowledgment and a little care.*
LESLIE ASHMAN, RESTON, VIRGINIA, PROJECT MANAGER AND GAIA CONSCIOUSNESS ADVOCATE

When I am with trees, I start to feel these little thoughts about them. I am drawn to the silhouettes, lines, angles, and colors that trees display before me. As I take my daily walk through the park nearby, I ask myself, "What lessons can I learn from the trees today?" And, a beautiful lesson always comes. I hear their messages in the wind.

ANN ST. GERMAINE, EATONTOWN, NEW JERSEY, EDUCATOR, WRITER AND PHOTOGRAPHER

TRY THIS AGAIN:
"STEP INSIDE THE PLANT'S WORLD" GUIDED VISUALIZATION EXPERIENCE

Basia says: We've included this exercise again because it is so important. Of course, you can do it any time and make notes about your experience in Part 6: Free-Writing or Journaling beginning on page 101.

Read through all the steps. Decide whether you feel comfortable doing this exercise. If you do not feel comfortable or have any concern whatsoever, do not do it.

Preparation: Make sure you are in a pleasant and private place where you won't be interrupted for about 15 minutes. There will be times during the exercise that you may want to close your eyes. If you are comfortable doing that, please do so.

• Make sure you have a plant near you that you can touch. You may also do this exercise standing or sitting outside privately with a tree.

• Write down some notes about your experiences. It's best to make notations immediately in case you don't remember later.

• Questions will be asked during the experience. Those questions are meant to be answered privately, inside yourself. You may jot notes for answers.

• Please remember this illustration as you do the exercise.

© Basia Alexander

Step 1: Begin.
❑ *Get comfortable.*
❑ *Breathe deeply and gently.*
❑ *Turn your attention to the plant.*
  Never mind its name. Forget anything that you may know about it.

Step 2: Since it is a living Being, always ask permission to make contact.
❑ *In your Heart, say to the plant or tree,*
  "I would like to spend some time with you and get to know you. Is that okay?"
❑ *You will probably feel a sense of calm. That means you have permission from the plant. It is unlikely, but if you feel disquiet, you may move to another plant, or stop.*

Step 3: Focus your sight on the stem, trunk, leaves, or flowers. Notice even the tiniest details.
❑ *See a multitude of shapes.*
❑ *Notice how the leaves are arranged on the stem.*
❑ *Notice how they are oriented in space.*
❑ *Distinguish shades of color.*
❑ *See how light or shadow plays around them.*

continues...

Step 4: Now, very gently, touch the plant or tree.

❏ *Feel the textures. Notice the shapes.*

❏ *Perceive curves and turns.*

❏ *Sense the temperature.*

Step 5: Use a soft focus and perceive the whole plant or tree.

Step 6: Ask yourself: **"How do I feel?"** and **"What do I know?"**
Quietly answer those questions for yourself or jot down a few notes on page 34.

Step 7: Focus for a moment on your own body. You may close your eyes.

❏ *Around your body, notice your heart's bioenergy field.*

❏ *Notice the bioenergy field's current size and shape.*

❏ *Imagine that your heart's biofield is like the sun—shimmering with light or radiating with energy.*

❏ *Imagine that the field is increasing in size and intensity.*

Step 8: Notice or imagine the bioenergy field of the plant or tree. You may close your eyes to do so.

❏ *Imagine that the plant's or tree's bioenergy field is also expanding. Feel it.*

❏ *Your heart's field now overlaps with the energy field of the plant or tree.*

❏ *In that overlapping area, information is shared and exchanged, sensory experience is stronger, and emotional perception is heightened.*

Step 9: With your eyes closed, for about a minute, allow yourself to be aware of any new information, experience, or perception in the bioenergy overlapping state (as the illustration on the facing page suggests).

Step 10: Step into the plant's or tree's world.

❏ *Engage your imagination and close your eyes.*

❏ *Feel as if you shrink or expand to fit the plant's or tree's size.*

❏ *Imagine that there is a door on the stem or trunk. You open it. A bright, white light shines on you, and you step inside the stem or trunk.*

❏ *There are thousands of cells all around you. Imagine reaching out with your hands and touching some cell walls.*

❏ *As you look around, you see little streams going up and down all around you. These are circulating plant fluids.*

❏ *Imagine that a tiny boat comes along. Hop in and begin moving upwards with the flow of the fluids. Continue moving upward.*

Step 11: Ride into a leaf.

❏ *You approach a leaf. The boat docks. You float forward, following a vein into the leaf.*

❏ *Sense light coming through the upper layers of cells.*

❏ *There is a lot of activity: A bubble of carbon dioxide is captured. A bubble of oxygen is released.*

❏ *Sense the heat from foods being produced.*

❏ *You are inside the leaf..*

Step 12: Ask yourself: **What do I sense?**   **What do I notice?**   **What do I realize?"** and **"What is important to the plant?"**
Quietly answer those questions for yourself or jot down a few notes on page 34.

The exercise continues on the next page...

Step 13: Ride the sugar molecule.

❑ *Inside the leaf, photosynthesis is producing sugar molecules, which are food for the plant. Imagine that you shrink down so small you can jump on a sugar molecule.*

❑ *Ride it into the stream. Travel out of the leaf and down the stem or trunk.*

Step 14: Move inside the roots.

❑ *Imagine that the stream flows beneath the soil and you gently submerge into the root zone. Imagine that–ahead–the little stream is splitting and narrowing.*

❑ *Your molecule stops on the side of the little stream as food.*

❑ *You step off the sugar molecule. You are inside the roots.*

Step 15: Ask yourself: **"What do I sense?   What do I notice?   What do I realize?"** and **"What is important to the plant?"**
Quietly answer those questions for yourself or jot down a few notes on page 34.

Step 16: Pick up the sponge.

❑ *Imagine turning around. Another stream is behind you. This one is flowing upwards.*

❑ *Now, imagine that you have a sponge. Step toward the stream, put in the sponge and sop up some water and nutrients. Another little boat comes along. Bring the sponge with you and get in.*

❑ *Feel a surge forward as the root's pumping action pushes you upward. You emerge and are traveling in the brightness, up the stem or trunk again.*

❑ *This stream flows toward a special part called the "growing point." Once there, you deposit the water and nutrient-soaked sponge in a growing cell.*

❑ *Imagine that the cell begins to divide and grow.*

Step 17: Fill the whole plant or tree.

❑ *Now, feel yourself expand and grow until you fill the whole plant or tree.*

❑ *Sense all the activity—thousands of things happening and interacting. All these interactions and feedback loops are its Growth Energy.*

❑ *Almost like a pulse or heartbeat, sense the strength of that Growth Energy.*

Step 18: Come from the plant's or tree's point of view.

❑ *Feel the movement of the Growth Energy around the plant or tree.*

❑ *Perceive where it is surging. Find out where it might be weak.*

Step 19: Ask yourself: **What do I sense?   What do I notice?   What do I realize?"** and **"What is important to the plant?"**
Quietly answer those questions for yourself or jot down a few notes on page 34.

Step 20: Enjoy a few moments with the tree or plant.

Step 21: BEING the tree or plant and the Flow of Collaboration

❑ *Realize that you have made a new friend and can do something like "see" through its "eyes" or "feel" through its life. Feel the two-way flow of connection between you and the tree or plant.*

❑ *Notice that you are coming from the plant's or tree's point of view and not coming so much from your own point of view.*

❑ *Acknowledge that you and the plant or tree have created a conscious partnership. Feel the collaboration between the two of you like a two-way street.*

Enjoy this experience for a few moments. Jot down notes on the next page.

continues...

_____

_____

_____

_____

_____

Step 22: One at a time—slowly and gently—ask these questions to the plant or tree and immediately write down any feelings or impressions you receive. Your unique perception may be to hear it, see it, or feel it.

❑ *Ask: "**Do you have an instruction for me?**" Jot any notes below.*

❑ *Ask: "**Do you have a message for me?**" Jot any notes below.*

❑ *Ask: "**Do you have a lesson for me?**" Jot any notes below.*

Step 23: Exchange gifts with the plant or tree.

❑ *Breathe in any gift(s) from the plant.*

❑ *Breathe out and say "**Thank You**" to the plant for letting you into its world.*

Step 24: Leave the plant or tree.

❑ *If you are still touching the plant, slowly release it.*

❑ *Return to your own body and to your own point of view. When you are ready, wiggle your toes. Take in a long, deep breath and stretch your back and shoulders.*

Step 25: If you have not already written anything, please write your answers to these questions below. If necessary, sit quietly with your pen poised on the paper. Try writing some pleasant words just to get the pen moving.

❑ *What was important to the plant?*

❑ *What were any messages, instructions, or lessons from the plant?*

❑ *What gift did you receive?*

Step 26: Share your experience with other people.

❑ *Think of a friend or family member with whom you may share your personal experience. Imagine sharing this experience with them.*

❑ *If you feel safe, actually tell them about it. If you don't feel safe or think it would be a bad idea, don't do it.*

_____

_____

_____

_____

_____

_____

continues...

continues...

Sometimes I feel that trees say funny things to me and that gives me great joy. On one occasion, I felt the presence of someone walking down the path and seeing me giggling with a tree. I said to myself, "Oh well, it's silly to be self-conscious." I realized that it didn't matter to me whether someone saw me or what they thought of me.
It didn't matter because I received so much joy from the tree. Feeling that joy overpowered any self-consciousness. It was greater than any judgment that I or anyone else might have of me. Then a thought came to me, "How wonderful it could be to have that person experience it, too." I realized later, the thought was actually a message from the tree.

LORI MYRICK, EAST WINDSOR, CONNECTICUT, ENERGY THERAPY PRACTITIONER

THREE MESSAGES FROM THE PLANT KINGDOM

1) We are alive. We are living Beings and valuable assets to people and to the planet. We want to be honored.

2) We are weak. Thus we are more susceptible to hurt from insect Beings and disease Beings.

3) Our health can be restored with people's help, but only if they come inside of our world, see our lives from our point of view, and heal our inner functions, our energy flows, and our song.

FOUR CATEGORIES OF MESSAGES FROM TREES AND PLANTS

1) *What's going on inside and outside for the Green Being's health.* This includes what is going on inside of its physiology and/or what the environmental conditions are that influence the tree or plant on either the inside or outside.

2) *Instructions to humans.* These are actions to take like "water, now" or "stop watering so much."

3) *Personal messages for you.* Get information that is useful in your personal life like "spend more time breathing deeply" or "bring your children to visit me."

4) *Insights or wisdom from Spirit;* timeless information or advice.

METHODS OF COMMUNICATION THROUGH WHICH PEOPLE CAN RECEIVE MESSAGES FROM NATURE'S CONSCIOUSNESS

• **Physical Sense Experiences**, such as tingling, warmth, visualizations, sounds, words, fragrances, etc.

• **Emotional Feelings**, such as a deep compassion, love, humor, or joyousness; also feelings of fear or thoughts that warn of dangers.

• **Direct Perception or Intuitive Perception**, such as inner hearing, inner seeing, inner knowing, etc.

• **Impressions** during spiritual practices or personal experiences of meditation, prayer, chanting, dreams, epiphany, etc.

• **Impulses** during artistry and/or creativity, such as inspiration during painting, sculpting, singing, dancing, writing, etc., as well as the **beauty of the result of that creativity.**

• **Messages arriving in meaningfully synchronous ways,** such as overheard conversations, songs on the radio, signs on trucks, books opening, sudden appearances of animals or insects, etc.

• **Various energy vibrations or light stimuli** that can be translated into the spoken or written word, or other expressions.

As a sensitive healer with people, I develop my perceptions with trees and plants in the simplest way I can. I try to feel like I am talking on the phone: a connection gets made and I hear the tree or plant speak.

ALANA DUBOIS, ROBBINSVILLE, NEW JERSEY, MASSAGE THERAPIST, REIKI AND ENERGY-MEDICINE INSTRUCTOR

Part 3: Messages from Trees and Plants

Dr. Jim reminds: Chapter 3 of our main book *Tree Whispering: A Nature Lover's Guide to Touching, Healing, and Communicating with Trees, Plants, and All of Nature* explains how trees and plants are alive, creative, intelligent and even spirited! Messages received by people from various trees and plants are quoted, too. Chapter 4 goes into detailed instructions about how *you* can be a better receiver of messages.

In the previous section of this notebook/journal, you did key exercises from the main book's Chapters 3 and 4 that prepare you to receive messages *yourself* from the Beings of Nature.

Basia talks about ways to receive messages: Now that you have advanced through many of the exercises in this notebook/journal, you already trust yourself and trust the path that you are on.

Please acknowledge that you are already a master at interpreting nonverbal signals from other people. Give yourself credit for understanding the intricacies of vocal inflection, body stance, and emotional expressions. If you are a parent, you grasp your baby's needs without words. If you have animals, well, they may as well be talking to you because you understand what they are "saying." Your skills are already sensitive and advanced. Consider that you can transfer those abilities into the realm of communication with the Plant Kingdom.

Messages come from Nature's Consciousness all the time and you probably take them in stride. The "Methods of Communication" box on the left outlines some ways you can be a receiver.

Throughout this notebook/journal you have read the nuggets of wisdom and advice about getting messages from trees and plants from graduates of our workshops and classes. They all echo the same principles: to have an open mind and expect that messages will come forth.

Dr. Jim says: The four basic categories of messages are shown to the left. Messages that you get may be short or long. The trees that talk to me tend to be concise, as well as to offer a wide variety of comments and insights about their lives. You may see or hear something personal about yourself or profound wisdom. You may simply have a good, warm feeling inside. That is enough!

As you are doing all these exercises and moving along the path, you will get messages from trees and plants.

This section of the notebook/journal is the place to write those messages down. It's *all* about collaboration and getting messages!

I think a lot of people are hesitant to believe they can receive messages from trees. They wonder, "Am I putting my own thoughts in?" I would advise you to slow down, allow your own Spirit and your energy some freedom. Believe in your energy and the path it may take. It will only give you more energy and freedom.
CHERYL SMITH, PhD, UNIVERSITY OF NEW HAMPSHIRE, EXTENSION PROFESSOR AND PLANT HEALTH SPECIALIST

As a child, if I was upset, I would go to my favorite tree and feel better. I also wrote poetry while sitting beneath its branches. I seemed to get phrases; ideas would just pop in. Now, I'm sure that the tree was helping me. If you have a favorite tree from childhood, I would recommend that you return to it. Perhaps you can reclaim some of your childhood wonder. It may be happy to see you again and give you fresh insights and inspiration just as my childhood tree did.

DEBBRA GILL, NEW YORK, NEW YORK, HOLISTIC NUTRITION AND
WELLNESS DIRECTOR FOR AN INTERNATIONAL CHILDREN'S ART PROJECT

I used to get clear messages as a young person. Now it seems that if a message doesn't show up on a flaming billboard, then I think I'm not getting it. But I know that's not true. So, I have to be aware that if I'm not getting a "wow," then either the tree doesn't communicate that way, or I'm not open. I have criticized myself for losing touch, but I have to realize that people change and, so, forgive myself.   Tchukki Andersen, Billerica, Massachusetts, arborist

Trees talk to me in the rustling of their leaves and movement of their branches, sending messages about themselves and the planet. In receiving their messages, I enter into partnership with them.
LESLIE ASHMAN, RESTON, VIRGINIA, PROJECT MANAGER AND GAIA CONSCIOUSNESS ADVOCATE

My girlfriend likes to hold a party once a year when her enormous Cherry tree blooms. During the most recent celebration, she asked all of us to touch the tree and get a message from it. Most of the people attending had not taken the Tree Whispering class; they reported that the tree was happy and grateful to her for her love. That was probably true; at the same time I got a message that was somewhat disturbing but certainly practical.

It said, "I am suffocating and need to breathe." I looked around and saw that my girlfriend's patio was made of bricks that were laid too close to the trunk of the Cherry tree, covering much of the root area. I told her that her tree needed to get air into its roots; it needed to have some of the bricks removed. Because of the class, I know that a tree can look great—such as having lots of blossoms—but the next year it can start to weaken.

CAROL HULLEY, KINGS PARK, NEW YORK, GARDENER

Often people's beliefs about what is possible create a barrier that doesn't let in anything they can't see, touch, or feel. So, people lose the connections to their imaginations and childlike wonder. It's okay to be like a child again—one who talks with plants, trees, butterflies, or bugs. That innocence is in everyone; it is going to connect to the purity of the Nature Being. Everyone can enjoy this connection through imagination.

MADELINE "GROWEESHA" THOMPSON, BOONTON, NEW JERSEY, BUSINESS OWNER AND PROFESSIONAL COUNSELOR

ᘓᔔ *A friend gave me a small red Azalea bush which I planted in my front yard. After a few years,*
   I noticed that it wasn't growing much. So, I talked with it, and it said it was lonely. It wanted a
companion. But, it specified that it did not want another red one. It gave me a clear picture of a
white Azalea and showed me—in my mind's eye—where to plant the new one, nearby. I marked the
spot with a stick and left to get the companion. It's been another year and both are growing happily.
                    GEORGETTE HRITZ, SCOTCH PLAINS, NEW JERSEY, POSTAL WORKER AND HOMEOWNER

I moved out of Manhattan many years ago. On a return trip to my old neighborhood to visit a friend, I walked past the small park adjacent to the Museum of Natural History. Suddenly, I felt as if a certain tree's heart jumped out to greet my heart. "It's good to see you again!" is what I felt this gnarly, old tree saying to me. I was stunned, but I did remember that tree from countless walks past it to the subway station. Upon reflecting, I realized that in my years away, I did a lot of global travel to study my craft and expand my spirituality. When I walked past the tree again, I was able to spontaneously receive Nature's intelligence in a whole new way.

MARISE HAMM, SAG HARBOR, NEW YORK, FENG SHUI CONSULTANT

There is a Copper Beech on the vast grounds of the Glastonbury Abbey with whom I have a strong affinity. Every year, when I arrive with my tour group, I can sense this Beech reaching out to me as soon as I get off the bus. When the Beech comes into sight, my heart leaps with delight. Tears of joy are usually streaming down my face. At last, I embrace the massive gray trunk and am enveloped by this presence. "Welcome home," it tells me.

CAROL OHMART-BEHAN, ENDICOTT, NEW YORK, AUTHOR AND GUIDE OF SPIRITUAL JOURNEYS

I was terribly upset at the Racine Parks Department for aggressively cutting down so many beautiful trees in the city's parkway. Due to a recent storm, many branches came down. I feel that the city used that storm as an excuse to cut down trees.
As I lay in bed at night, angered by this tree slaughtering, I sensed the trees whispering to me, "Do not be angry. Your anger doesn't do us any good. What is to be, will be. Would you take that anger and turn it into something positive that would benefit and bless us?"
MARY CYPRESS, RACINE, WISCONSIN, STREET ARTIST

We don't mind too much that the humans cut us back into shapes and sizes. It is like a haircut for us. Overall, the humans are caring to us, as best they can from their point of view. However, if the humans would ask us how we should be shaped and cut—if they would come from our point of view—we would be far more beautiful than anything the humans could imagine.

PLANTS IN A SEATTLE, WASHINGTON, ORNAMENTAL GARDEN

*In preparing readers to receive our messages, remind them
that we also have a sense of humor.*
THE DESIGN INTELLIGENCE OF THE PLANT KINGDOM

THE FIVE HERALDS (SEE CHAPTERS 7 & 8 IN THE MAIN BOOK)

BE THE TREE

Ask for a bioenergy connection with another life form. Settle into the overlapping connection with it in a sensory and emotional way. By experiencing the plant's way of life, you are BEING the Green Being. A harmony occurs called "coming from the tree's point of view." This step is transformational for both you and the tree or plant.

ASK THE TREE

It's only respectful to ask. Asking leads to partnership. It's like playing the game "20 Questions." Feel the Green Being's "yes" or "no" responses within your body or intuition. Let a "no" response be a guide to lead you to a better way by asking more questions.

The more knowledge you have, the better questions you can ask. By asking the Green Being specific questions, you strengthen and/or expand your connection with it; you get more information about it.

HEAL THE TREE

The Green Being's internal functionality needs to be restored first. The bioenergy interactions between a person's healing intent and the Green Being's natural drive to be healthy result in initiating the healing process. Techniques include the seven Healing Whispers™. The act of healing trees and plants may be a healing experience for people, too.

SAVE THE TREE

Healing of internal functionality drives growth. Growth drives healing of internal functionality.

Healing ⟶ Growth

Feel the shift of bioenergy—from the pulling in of decline shifting to the pushing out toward growth. The tree is in the process of being SAVED.

Be patient. Be observant. Give the Green Being the gift of plenty of time to show new growth.

LOVE THE TREE

Love is always there. It does not come last in this process. When a person comes in contact with the Life Force and bioenergy of another Being, there is a beautiful experience of purity. At the point of connectedness, the balance of purity flows back and forth. That purity flowing back and forth can only be described as LOVE.

Part 4: The Five Heralds
Being, Asking, Healing, Saving, Loving

The Five Heralds
- **BE the Tree**
- ASK the Tree
- HEAL the Tree
- SAVE the Tree
- LOVE the Tree

TRY THIS: BE THE TREE—CREATE A BREATHING CONNECTION CYCLE
Read through these steps and decide whether you feel comfortable doing the exercise. If you do not feel comfortable or have any concern whatsoever, do not do it.
Audios are available to download at www.TreeWhispering.com.

Step 1: Go to a specific tree. Describe which tree: _____

_____

Step 2: Put your hands on the tree. Close your eyes, if comfortable doing so.

Step 3: Imagine a stream of air coming into your heart area from the tree and going out your belly area back to the tree. This stream of air creates a connection cycle.

Step 4: On the incoming stream of air, imagine the Life Force and bioenergy of the tree connecting with you.

Step 5: On the outgoing stream of air, imagine your Life Force and bioenergy connecting with the tree.

Step 6: Continue to make the cycle of connection: the tree's Life Force coming in at your heart area and your Life Force going to the tree from your belly area.

Step 7: Breathe in and out easily and gently. Allow yourself to enjoy this connection cycle for several minutes.

Step 8: Acknowledge that your connection allows you to BE THE TREE.

Step 9: Thank the tree for allowing you into its world. Jot down any notes. Writing a few brief phrases is okay. Let the words flow from your heart.

_____

_____

_____

_____

_____

_____

_____

_____

_____

_____

_____

The Five Heralds
(BE the Tree)
ASK the Tree
HEAL the Tree
SAVE the Tree
LOVE the Tree

TRY THIS AGAIN:
BE THE TREE—CREATE A BREATHING CONNECTION CYCLE

Read through these steps and decide whether you feel comfortable doing the exercise. If you do not feel comfortable or have any concern whatsoever, do not do it.

Step 1: Go to a specific tree. Describe which tree: _____

Step 2: Put your hands on the tree. Close your eyes, if comfortable doing so.

Step 3: Imagine a stream of air coming into your heart area from the tree and going out your belly area back to the tree. This stream of air creates a connection cycle.

Step 4: On the incoming stream of air, imagine the Life Force and bioenergy of the tree connecting with you.

Step 5: On the outgoing stream of air, imagine your Life Force and bioenergy connecting with the tree.

Step 6: Continue to make the cycle of connection: the tree's Life Force coming in at your heart area and your Life Force going to the tree from your belly area.

Step 7: Breathe in and out easily and gently. Allow yourself to enjoy this connection cycle for several minutes.

Step 8: Acknowledge that your connection allows you to BE THE TREE.

Step 9: Thank the tree for allowing you into its world. If necessary, sit quietly with your pen poised on the paper. Try writing some pleasant words just to get the pen moving. Once you are writing the first words, more words will follow.

continues...

It's hard on the trees when they can't be touched. If you are with a tree that can't be touched, take time to share your love with it energetically. Sharing breath is the best starting point. Be as quiet in the mind as possible and open your heart. Let go of the desire for anything in particular to happen. You can imagine a figure-8 or an infinity sign of breath and energy going back and forth between you and the tree.
ROBIN ROSE BENNETT, HEWITT, NEW JERSEY, AUTHOR, VISIONARY HERBALIST, AND RENOWNED TEACHER

The Five Heralds
BE the Tree
ASK the Tree
HEAL the Tree
SAVE the Tree
LOVE the Tree

TRY THIS: BE THE TREE—TRANSMITTING AND RECEIVING CYCLE

Read through these steps and decide whether you feel comfortable doing the exercise. If you do not feel comfortable or have any concern whatsoever, do not do it.

Audios are available to download at www.TreeWhispering.com.

Step 1: Go to a specific tree. Describe which tree: _____

_____

Step 2: Put your hands on the tree. Close your eyes, if comfortable doing so.

Step 3: Imagine that the tree is a transmitting tower. Imagine that it is sending out something like radio waves in all directions.

Step 4: Imagine that you are a radio receiver. Imagine that you can tune in and receive the tree's waves.

Step 5: Imagine the reverse. You are a transmitting tower and the tree is a radio receiver, receiving your waves.

Step 6: Imagine that the tree and you are making a circular, two-way, transmitting-and-receiving cycle.

Step 7: Breathe in and out easily and gently. Allow yourself to enjoy this two-way, transmitting-and-receiving cycle for several minutes.

Step 8: Acknowledge that this connection allows you to BE THE TREE.

Step 9: Thank the tree for allowing you into its world. Jot down any notes. Once you are writing the first words, more words will follow.

_____

_____

_____

_____

_____

_____

_____

_____

_____

_____

_____

_____

_____

The Five Heralds
(**BE the Tree**)
ASK the Tree
HEAL the Tree
SAVE the Tree
LOVE the Tree

Try This AGAIN:
BE THE TREE—Transmitting and Receiving Cycle

Read through these steps and decide whether you feel comfortable doing the exercise. If you do not feel comfortable or have any concern whatsoever, do not do it.

Step 1: Go to a specific tree. Describe which tree: _____

_____

Step 2: Put your hands on the tree. Close your eyes, if comfortable doing so.

Step 3: Imagine that the tree is a transmitting tower. Imagine that it is sending out something like radio waves in all directions.

Step 4: Imagine that you are a radio receiver. Imagine that you can tune in and receive the tree's waves.

Step 5: Imagine the reverse. You are a transmitting tower and the tree is a radio receiver, receiving your waves.

Step 6: Imagine that the tree and you are making a circular, two-way, transmitting-and-receiving cycle.

Step 7: Breathe in and out easily and gently. Allow yourself to enjoy this two-way, transmitting-and-receiving cycle for several minutes.

Step 8: Acknowledge that this connection allows you to BE THE TREE.

Step 9: Thank the tree for allowing you into its world. Write down any notes.

_____

_____

_____

_____

_____

_____

_____

_____

_____

_____

When I am BEING the tree, I get a better connection with the tree. Then, whatever I feel informs my actions. I remember what Dr. Jim said: "Forget everything. Forget the botany and the physiology that you know. Forget all the learning and just listen to the tree." To me, forgetting equates to humility. Humility is tough for me because I am proud of what I know. But, I come from the tree's point of view to hear from the tree, not me.

MIKE NADEAU, SHERMAN, CONNECTICUT, HOLISTIC LAND-CARE PRACTITIONER

The Five Heralds
BE the Tree
ASK the Tree
HEAL the Tree
SAVE the Tree
LOVE the Tree

Try This: BE THE TREE—Put Yourself in the Tree's Place

Read through these steps and decide whether you feel comfortable doing the exercise. If you do not feel comfortable or have any concern whatsoever, do not do it.

Audios are available to download at www.TreeWhispering.com.

Step 1: Go to a specific tree. Describe which tree: _____

Step 2: Put your hands on the tree. Take your shoes and socks off and/or close your eyes, if comfortable doing so.

Step 3: Breathe gently and easily for about a minute while holding the tree.

Step 4: Imagine your feet sinking into the earth and growing roots for a minute.

Step 5: Imagine standing with your arms stretched up toward the sun. Imagine your fingers growing leaves and absorbing the heat and energy of the sun. Do this for a minute.

Step 6: Imagine the sun on your face, the breeze in your hair, and rainwater soaking up from the earth. Do this for a minute.

Step 7: Breathe in and out easily and gently. Allow yourself to enjoy this connection for a few minutes.

Step 8: Acknowledge that this connection allows you to BE THE TREE.

Step 9: Thank the tree for allowing you into its world. Write down any notes. Try writing pleasant words just to get the pen moving.

The Five Heralds
BE the Tree
ASK the Tree
HEAL the Tree
SAVE the Tree
LOVE the Tree

TRY THIS AGAIN:
BE THE TREE—PUT YOURSELF IN THE TREE'S PLACE

Read through these steps and decide whether you feel comfortable doing the exercise. If you do not feel comfortable or have any concern whatsoever, do not do it.

Step 1: Go to a specific tree. Describe which tree: _____

_____

Step 2: Put your hands on the tree. Take your shoes and socks off and/or close your eyes, if comfortable doing so.

Step 3: Breathe gently and easily for about a minute while holding the tree.

Step 4: Imagine your feet sinking into the earth and growing roots for a minute.

Step 5: Imagine standing with your arms stretched up toward the sun. Imagine your fingers growing leaves and absorbing the heat and energy of the sun. Do this for a minute.

Step 6: Imagine the sun on your face, the breeze in your hair, and rainwater soaking up from the earth. Do this for a minute.

Step 7: Breathe in and out easily and gently. Allow yourself to enjoy this connection for a few minutes.

Step 8: Acknowledge that this connection allows you to BE THE TREE.

Step 9: Thank the tree for allowing you into its world. Write down any notes.

_____

_____

_____

_____

_____

_____

_____

_____

_____

_____

_____

_____

Dr. Jim says we should come from the tree's point of view. To me, this means "being" the tree and finding out from the tree what it is going through in its life.
I used to think I was a good arborist—figuring out what I should do for a tree. I thought I had all the answers. But now, I realize I can ask the trees and they tell me about their health. Now, I can do an even better job as an arborist.   PAUL O'KULA, LONG ISLAND, NEW YORK, ARBORIST, BUSINESS OWNER

The Five Heralds
BE the Tree
(ASK the Tree)
HEAL the Tree
SAVE the Tree
LOVE the Tree

TRY THIS:
RECEIVE "TRUE" OR "FALSE" RESPONSES FROM YOUR OWN BODY

Read through the steps. Decide whether you feel comfortable doing this exercise. If you do not feel comfortable or have any concern whatsoever, do not do it.
Audios are available for download at www.TreeWhispering.com.

Step 1: Be in a pleasant and private place where you won't be interrupted for about 10 minutes.

Step 2: Stand up, if convenient in your location. Take some deep breaths.

Step 3: Experience "true" or "false" responses in your body.
- ❑ *Read the following statements one at a time. Speak each one aloud.*
- ❑ *Pause between each statement and allow yourself to feel or sense your body's "true" or "false" response. Then, make the next statement.*

Note: Each statement is designed to evoke a simple "true" or "false" response, not a psychological inquiry.

❑ *I am an American.*	❑ *I am in my 50s.*
❑ *I am a Brazilian.*	❑ *I own a car.*
❑ *My favorite color is blue.*	❑ *I own a bicycle.*
❑ *My favorite color is green.*	❑ *My name is Mary.*
❑ *My favorite color is yellow.*	❑ *My name is Joe.*
❑ *I am a son.*	❑ *I live in an apartment.*
❑ *I am a daughter.*	❑ *I live in a house.*
❑ *I am in my 30s.*	❑ *I am wearing shoes right now.*
❑ *I am in my 40s.*	❑ *I am wearing clothes right now.*

Step 4: Address your body with the following requests:
- ❑ *"Body, please show me a 'yes.'"*
- ❑ *"Body, please show me a 'no.'"*

Step 5: Address your body with the following requests:
- ❑ *"Body, please show me a bigger, clearer 'yes.'"*
- ❑ *"Body, please show me a bigger, clearer 'no.'"*

Step 6: You may sit down, if you were standing. Write down any notes about your "true/false" or "yes/no" response experience.

_____

_____

_____

_____

_____

_____*continues...*

We are accustomed to speaking a verbal language. However, our bodies have their own language. Science has shown that our nervous systems respond to stimuli even faster than our minds. Remember the advice you got in school for "true/false" tests? It probably was: "Always take your first answer." I believe that is because you can feel a "yes" or "no" response within your body when you make a true or false statement.   BASIA ALEXANDER, THE CHIEF LISTENER

The Five Heralds
BE the Tree
ASK the Tree
HEAL the Tree
SAVE the Tree
LOVE the Tree

TRY THIS:
RECEIVE "YES" OR "NO" RESPONSES FROM A GREEN BEING

Read through the steps. Decide whether you feel comfortable doing this exercise. If you do not feel comfortable or have any concern whatsoever, do not do it.

Note 1: You will need to be able to touch a plant or a tree for this exercise.

Note 2: Since you will be in a bioenergy overlap with the plant or tree, you can get its "yes" and "no" responses or answers through your own nervous system by accessing the same or similar "true" or "false" feelings you had in your body or intuition previously.

Step 1: You may either sit or stand.

Step 2: BE THE TREE or BE THE PLANT as you did in exercises from earlier in this notebook/journal. Touch the tree or plant. Create a two-way flow of Life Force and bioenergy connection with the tree or plant. You may want to review the "Stepping Inside the Plant's World" exercise.

Step 3: Open yourself to a pleasurable experience of communication with a tree or plant. Go to a specific tree. Describe which tree: _____

_____

Step 4: Ask permission by saying in your heart or aloud, "**I ask for permission to interact with you in this exercise.**"

❑ *If you feel disquiet or agitation, please stop. You do not have permission.*

❑ *If you feel calm or peaceful, continue here.*

Step 5: Say in your heart or aloud, "**I ask to connect with the earth, the sky, the Consciousness of Nature, and this Green Being's Growth Energy. Nature Consciousness, for the highest good and in a gentle way, please assist me with a heightened intuitive, sensory, and emotional connection with this plant.**"

❑ *Notice that your heart's biofield and the bioenergy field of the tree or plant are overlapping.*

❑ *If you feel disquiet or agitation, stop. If you feel calm or peaceful, continue here.*

Step 6: Ask the Green Being, "**Please show me a 'yes,'**" and feel a response.

Step 7: Ask the Green Being, "**Please show me a 'no,'**" and feel a response.

Note: If the plant's responses are unclear at this point, be patient and continue here.

Step 8: Slowly and patiently, ask the plant each of the following questions. Receive its "yes" or "no" response to each question before moving on to the next question.

Note: Remember that "yes" or "no" are the only desired responses in this exercise.

a) "Do you have sufficient water available?"

b) "Have you been transplanted?"

c) "Are you connected with the deep earth energies?"

d) "Are you operating at 100% functionality?"

e) "Would you like to receive a Healing Whisper?"

f) "Do you have a message for me?"

continues...

Step 9: Ask the plant, "**Please show me a big 'yes,'**" and feel a response.

Step 10: Ask the plant, "**Please show me a big 'no,'**" and feel a response.

Step 11: Say, "**thank you**" to the plant for letting you into its world.

Step 12: Write notes about your "yes" and "no" experience with the plant.

Note: You may practice this exercise with several Green Beings.

I practice and teach a healing system that is fundamentally based upon using our innate ability to receive information about a client's bodymind complex. We honor that Being by always asking for permission prior to performing a session.

As an instructor, I show others how to hone their intuitive skills to receive "yes" and "no" answers when working with a client to determine which body/mind areas are priorities to focus on.

In a class, a student was asking for permission to treat a fellow student, but kept getting "no" answers. A couple of us did some detective work to find out where the "no" was coming from. We determined that the student was picking up the "no" answer from a nearby plant because it needed care. After all, we are all connected! Applying the same intuitive skills, a questioning process, and using "yes" and "no" feedback, we discovered that it wanted the trash in its pot removed, to be in between two other plants its same size, and to be far enough from the heat vent. We complied with its needs. Then—and only then—did the student get a "yes" to continue with practice.

MELANIE BUZEK, CORNVILLE, ARIZONA, PHYSICAL THERAPIST, ENERGY-MEDICINE PRACTITIONER AND EDUCATOR

The Five Heralds
BE the Tree
(ASK the Tree)
HEAL the Tree
SAVE the Tree
LOVE the Tree

TRY THIS: ASK THE TREE OR PLANT

Read these questions first. Before you attempt to ASK THE TREE, always take care of yourself and make sure that you feel comfortable. If you have any concerns, do not do this.

Here is a list of yes/no questions that you can use when you want to get information from a tree, plant, or other Being of Nature. Use the yes/no question and answer process taught in the previous exercise as a guidance system.

Adapt this list of questions to fit your particular situation. When in doubt, or if you don't understand, continue to ask your own yes/no questions until the answer is clear to you.

By using this list, you are giving the tree or plant a chance to say "no." "No" may not mean "no, never." A "no" answer may indicate that the plant needs you to ask different questions.

Step 1: Go to a specific tree or plant. Describe which one: _____

_____

Step 2: BE THE TREE. Create a two-way flowing Life Force and bioenergy connection with the tree or plant.

Step 3: Ask questions as appropriate to your situation:

❑ **"Do I have permission to ask you some questions?"** _____

❑ **"Do you need water?"** _____

    ○ *1 cup?* ○ *1 pint?* ○ *1 quart?* ○ *1 gallon?* ○ *drip hose for 2 hours?* _____

    ○ *more?* ○ *less?* _____

❑ **"Do you need a tie, band, or wire removed?"** _____

❑ **"Do you need organic compost?"** _____

    ○ *now?* ○ *later?* ○ *when?* _____

    ○ *how much of a specific compost? or what kind?* _____

    ○ *more?* ○ *less?* _____

❑ **"Do you need food?"** _____

    ○ *now?* ○ *later?* ○ *when?* _____

    ○ *how much of a specific food—such as sea kelp? or what kind?* _____

    ○ *more?* ○ *less?* _____

❑ **"Do you need beneficial microorganisms?"** _____

    ○ *now?* ○ *later?* ○ *when?* _____

    ○ *how much of a specific product? or what kind?* _____

    ○ *more?* ○ *less?* _____

    ○ *on leaves?* ○ *in soil around roots?* _____

❑ **"Do you need a dead limb removed?"** _____

    ○ *now?* ○ *later?* ○ *when?* _____

continues...

❏ **"Do you need the soil under you loosened up?"** _____

  ○ *now?* ○ *later?* ○ *when?* _____

  ○ *how much--lightly?* ○ *how much--more vigorous?* _____

❏ **"Is a certain pest a problem for you?"** _____
  (Look for one you might see on or around the plant.)

  ○ *does it need to be removed?* _____

  ○ *now?* ○ *later?* ○ *when?* _____

  ○ *use a _____ product? (be specific)* _____

  ○ *is there another way to handle this pest problem? (ask about specific options)* __

  _____

  ○ *If a product, ask how much? (always act within product label direction guidelines)*

  _____

❏ **"Is this a good transplant location for you?"** _____

  ○ *too much sun?* ○ *not enough sun?* _____

  ○ *location?* _____ ○ *orientation?* _____

❏ **"Do you need to be left alone?"** _____

❏ **"Do you need to be loved or appreciated?"** _____

Step 4: Take action in cooperation with the tree or plant, coming from its point of view. Always ask more questions if you are unclear about any step in the process.

_____

_____

Step 5: Thank the tree or plant for allowing you into its world. Write notes here.

_____

_____

_____

_____

_____

_____

For years, I have been listening to trees and plants, learning how to come from their points of view. To do that, I have had to check my ego at the door. In other words, I have had to set aside my own personal point of view. I have learned to ask the plants and trees what is best for them and to have humility in doing so.   DR. JIM CONROY, THE TREE WHISPERER

A Gift for You:
The First Healing Whisper

HEALING WHISPERS™
Experiential Bioenergy Healing Techniques
Done In Partnership and Cooperation
with Trees and Plants

Dr. Jim says: There are seven Healing Whispers™. Here is a gift of the first Healing Whisper for you!

First, try making up your own questions and engage the Green Being in an exchange about what is going on inside of it, as you have done previously. With your resulting deeper understanding, the Healing Whisper will be able to imprint a healthy pathway inside of the plant.

The Five Herald
BE the Tree
ASK the Tree
(HEAL the Tree)
SAVE the Tree
LOVE the Tree

TRY THIS: HEAL THE TREE—PERFORM THE HEALING WHISPER
Read through the steps. Decide whether you feel comfortable doing this exercise. If you do not feel comfortable or have any concern whatsoever, do not do it.
Audios are available to download at www.TreeWhispering.com.

Review: Consider returning to the "Stepping Inside Their World" guided meditation exercise.

Step 1: Go to the tree or plant that you think needs the Healing Whisper. Describe its appearance now: _____

_____

Location: _____

Step 2: Sit comfortably or stand with the tree or plant for a few minutes.

Step 3: Ask for permission to connect in the following way:
- *Breathe slowly and deeply. Remove shoes and/or close your eyes, if possible.*
- *Imagine the size and shape of your own heart's biofield.*
- *Imagine the tree's or plant's bioenergy field.*
- *Ask for permission:* **"May I connect with you?"**

If you feel peaceful or get a "yes," then you have permission.

If you get a "no" or feel disquiet or agitation, then you don't. Drink some water. Check on your own level of attentiveness. Then, try asking again.
- *Imagine a connection with the Green Being or experience overlapping of the two biofields.*
- *Allow yourself to enjoy, see, hear, smell, sense, feel, or intuit.*
- *Relax into coming from the Green Being's point of view. You may receive information or thoughts. You may feel sensations.*

Step 4: In your heart, ask the Green Being for permission to interact. **"I ask permission to give you a Healing Whisper."**
- *If you feel peaceful or get a "yes," then you have permission.*
- *If you get a "no" or feel disquiet or agitation, then you don't. Drink some water. Check on your own level of attentiveness. Then, try asking again.*
- *If you get a "no" again, honor it. **Stop** the interaction. Thank the plant.*

continues...

Step 5: If you received a "yes" in Step 4, hold the Green Being with one hand. With the other hand, touch or tap on the roots first and then on the stem. Repeat 4 to 8 times aloud or in your heart, while tapping:
"Nature Consciousness: Please remove blockages and distribute this Green Being's Growth Energy where it is needed."

Step 6: Check in with your intuition. Find out whether the Green Being fully received the treatment. Say, "**Did you receive the Healing Whisper?**"

❑ *If you have an awareness that the Green Being did not receive the treatment or if you get a "no" answer, repeat Step 5.*

❑ *If you have an awareness that it did get the Healing Whisper or you get a "yes" answer, continue here.*

Step 7: In your Heart, ask the Green Being: **"Do you have a message for me?"**

❑ *Using a soft focus, allow words or images, sounds, or symbols to come to you.*

❑ *Accept whatever comes, even if you think it is nothing. Some feeling or thought will come to you.*

Step 8: Say in your Heart: **"Thank you for letting me into your world."**

❑ *Ask: "***How long until the Healing Whisper is needed again?***"*

❑ *Feel gratitude. If you are still touching the Green Being, release your hold.*

Step 9: Be open to perceiving a sign of confirmation such as sudden wind blowing, appearance of something, a scent from the Green Being, etc.

Step 10: Write notes below about your experience.

Step 11: If you received a message, take special care to write that down below, especially if the tree or plant wants additional care.

Step 12: Find the right person with whom to share your experience and/or to share the message from the Green Being (especially about additional care).

Both our bodies and the trees are wise guides. They teach us how to grow through our challenges and be fully connected to the Source.
MADELINE *"GROWEESHA"* THOMPSON, BOONTON, NEW JERSEY, BUSINESS OWNER & PROFESSIONAL COUNSELOR

TRY THIS AGAIN:

HEAL THE TREE—PERFORM THE HEALING WHISPER

Read through the steps. Decide whether you feel comfortable doing this exercise. If you do not feel comfortable or have any concern whatsoever, do not do it.

Audios are available to download at www.TreeWhispering.com.

Review: Consider returning to the "Stepping Inside Their World" guided meditation exercise.

Step 1: Go to the tree or plant that you think needs the Healing Whisper.

Describe its appearance now: _____

_____

Location: _____

Step 2: Sit comfortably or stand with the tree or plant for a few minutes.

Step 3: Ask for permission to connect in the following way:

❑ *Breathe slowly and deeply. Remove shoes and/or close your eyes, if possible.*

❑ *Imagine the size and shape of your own heart's biofield.*

❑ *Imagine the tree's or plant's bioenergy field.*

❑ *Ask for permission:* **"May I connect with you?"**

If you feel peaceful or get a "yes," then you have permission.

If you get a "no" or feel disquiet or agitation, then you don't. Drink some water. Check on your own level of attentiveness. Then, try asking again.

❑ *Imagine a connection with the Green Being or experience overlapping of the two biofields.*

❑ *Allow yourself to enjoy, see, hear, smell, sense, feel, or intuit.*

❑ *Relax into coming from the Green Being's point of view. You may receive information or thoughts. You may feel sensations.*

Step 4: In your heart, ask the Green Being for permission to interact.

"I ask permission to give you a Healing Whisper."

❑ *If you feel peaceful or get a "yes," then you have permission.*

❑ *If you get a "no" or feel disquiet or agitation, then you don't. Drink some water. Check on your own level of attentiveness. Then, try asking again.*

❑ *If you get a "no" again, honor it.* **Stop** *the interaction. Thank the plant.*

Step 5: If you received a "yes" in Step 4, hold the Green Being with one hand. With the other hand, touch or tap on the roots first and then on the stem. Repeat 4 to 8 times aloud or in your heart, while tapping:

"Nature Consciousness: Please remove blockages and distribute this Green Being's Growth Energy where it is needed."

continues...

Step 6: Check in with your intuition. Find out whether the Green Being fully received the treatment. Say, **"Did you receive the Healing Whisper?"**

❑ *If you have an awareness that the Green Being did not receive the treatment or if you get a "no" answer, repeat Step 5.*

❑ *If you have an awareness that it did get the Healing Whisper or you get a "yes" answer, continue here.*

Step 7: In your Heart, ask the Green Being: **"Do you have a message for me?"**

❑ *Using a soft focus, allow words or images, sounds, or symbols to come to you.*

❑ *Accept whatever comes, even if you think it is nothing. Some feeling or thought will come to you.*

Step 8: Say in your Heart: **"Thank you for letting me into your world."**

❑ *Ask: "How long until the Healing Whisper is needed again?"*

❑ *Feel gratitude. If you are still touching the Green Being, release your hold.*

Step 9: Be open to perceiving a sign of confirmation such as sudden wind blowing, appearance of something, a scent from the Green Being, etc.

Step 10: Write notes below about your experience.

Step 11: If you received a message, take special care to write that down below, especially if the tree or plant wants additional care.

Step 12: Find the right person with whom to share your experience and/or to share the message from the Green Being (especially about additional care).

I walk regularly through Madison Park in New York City, where I live, and have a close relationship with a grandfather Elm and a grandmother Elm there. I went to the park many times all summer to do the Healing Whispers and other Tree Whispering techniques with the Elms. The next spring, they started to sprout out incredible growth. They almost looked like a bush—they had pushed out so many new little branches and leaves! I hope—because they look so good now—that the city parks department won't cut them down or trim them back. I feel that the trees are grateful to me for being their partner and being respectful to them.

DEBBRA GILL, NEW YORK, NEW YORK, HOLISTIC NUTRITION AND WELLNESS
DIRECTOR FOR AN INTERNATIONAL CHILDREN'S ART PROJECT

The Five Heralds
BE the Tree
ASK the Tree
HEAL the Tree
SAVE the Tree
LOVE the Tree

TRY THIS A THIRD TIME:

HEAL THE TREE—PERFORM THE HEALING WHISPER

Read through the steps. Decide whether you feel comfortable doing this exercise. If you do not feel comfortable or have any concern whatsoever, do not do it.

Audios are available to download at www.TreeWhispering.com.

Review: Consider returning to the "Stepping Inside Their World" guided meditation exercise.

Step 1: Go to the tree or plant that you think needs the Healing Whisper.

Describe its appearance now: _____

_____

Location: _____

Step 2: Sit comfortably or stand with the tree or plant for a few minutes.

Step 3: Ask for permission to connect in the following way:

❑ *Breathe slowly and deeply. Remove shoes and/or close your eyes, if possible.*

❑ *Imagine the size and shape of your own heart's biofield.*

❑ *Imagine the tree's or plant's bioenergy field.*

❑ *Ask for permission:* **"May I connect with you?"**

If you feel peaceful or get a "yes," then you have permission.

If you get a "no" or feel disquiet or agitation, then you don't. Drink some water. Check on your own level of attentiveness. Then, try asking again.

❑ *Imagine a connection with the Green Being or experience overlapping of the two biofields.*

❑ *Allow yourself to enjoy, see, hear, smell, sense, feel, or intuit.*

❑ *Relax into coming from the Green Being's point of view. You may receive information or thoughts. You may feel sensations.*

Step 4: In your heart, ask the Green Being for permission to interact.

"I ask permission to give you a Healing Whisper."

❑ *If you feel peaceful or get a "yes," then you have permission.*

❑ *If you get a "no" or feel disquiet or agitation, then you don't. Drink some water. Check on your own level of attentiveness. Then, try asking again.*

❑ *If you get a "no" again, honor it.* **Stop** *the interaction. Thank the plant.*

Step 5: If you received a "yes" in Step 4, hold the Green Being with one hand. With the other hand, touch or tap on the roots first and then on the stem. Repeat 4 to 8 times aloud or in your heart, while tapping:

"Nature Consciousness: Please remove blockages and distribute this Green Being's Growth Energy where it is needed."

continues...

Step 6: Check in with your intuition. Find out whether the Green Being fully received the treatment. Say, **"Did you receive the Healing Whisper?"**

❑ *If you have an awareness that the Green Being did not receive the treatment or if you get a "no" answer, repeat Step 5.*

❑ *If you have an awareness that it did get the Healing Whisper or you get a "yes" answer, continue here.*

Step 7: In your Heart, ask the Green Being: **"Do you have a message for me?"**

❑ *Using a soft focus, allow words or images, sounds, or symbols to come to you.*

❑ *Accept whatever comes, even if you think it is nothing. Some feeling or thought will come to you.*

Step 8: Say in your Heart: **"Thank you for letting me into your world."**

❑ *Ask: "How long until the Healing Whisper is needed again?"*

❑ *Feel gratitude. If you are still touching the Green Being, release your hold.*

Step 9: Be open to perceiving a sign of confirmation such as sudden wind blowing, appearance of something, a scent from the Green Being, etc.

Step 10: Write notes below about your experience.

Step 11: If you received a message, take special care to write that down below, especially if the tree or plant wants additional care.

Step 12: Find the right person with whom to share your experience and/or to share the message from the Green Being (especially about additional care).

𝒟 *In truth—I am really the "pass-through" for the healing to occur. I do not feel that I do the healing; I feel that the healing occurs through me.*
No healer can really explain the full extent of what he or she does. Usually, it is because the healer—in his or her ego personality—is not present. It seems paradoxical, but people who call themselves healers are usually operating as the vehicles for a metaphysical or spiritually-based occurrence. When I am treating a tree, my consciousness is deeply engaged in concentrating on details, but I feel as if "I" step aside and become a pass-through so that the true healing can occur. I am not giving the tree vibrations or energy as is done in certain human healing systems. Likewise, the tree is not giving energy to me.   DR. JIM CONROY, THE TREE WHISPERER

The Five Heralds
BE the Tree
ASK the Tree
HEAL the Tree
(SAVE the Tree)
LOVE the Tree

TRY THIS: SAVE THE TREE OR PLANT

Read through the steps. Decide whether you feel comfortable doing this exercise. If you do not feel comfortable or have any concern whatsoever, do not do it.

Note: Saving the tree or plant begins after you have given it the Healing Whisper. A shift of its bioenergy—from pulling in to pushing out—may occur. Healing drives growth. Growth drives healing. Be patient. Give the tree a chance to heal.

Return to a tree often that received the Healing Whisper.

At all times: Be patient. Maintain a caring connection with the tree or plant. Inspect it regularly and carefully. Take pictures.

Watch for the tiniest signs of healing or new growth. Things may happen such as leaking may start to dry up. Tiny, new leaves may appear.

Step 1: Check your notes on its appearance when you did the Healing Whisper. Write down notes about its current appearance and progress below.

Step 2: Put your hands on the Green Being and BE THE TREE. Use one of the imagining or breathing techniques from earlier in this notebook/journal.

Step 3: ASK THE TREE, **"Are you continuing to heal?"** _____

_____

Step 4: ASK THE TREE, **"Do you want the Healing Whisper again?"**
❑ *If "yes," ask when, and do it now, or whenever it says to do it. See pages 64, 66, or 68.*

Step 5: ASK THE TREE, **"Is there something else you need?"**
❑ *If "yes," go through the list in the ASK THE TREE section of this chapter. Emphasize items such as adding organic compost. Make notes about what it needs below.*
❑ *If "no," ask if it wants to be left alone while it continues to heal.*

Step 6: ASK THE TREE for some intuitive insights about what is going on for it. Be patient. The intuitive insight may take some time to come to you. Write down your intuitive impressions below.

Note: This step will give you some clarity as to what additional questions you might ask, or about doing the Healing Whisper, or if you should leave it alone while it continues to heal.

Step 7: ASK THE TREE if it needs some attention other than what you are capable of providing—including professional attention. ASK THE TREE if you should contact Dr. Jim at www.TheTreeWhisperer.com or a qualified local arborist.

Step 8: Thank the tree for letting you into its world. You may also write your comments, reflections, impressions, and notes here.

_____

_____

_____

_____

_____

_____*continues...*

All the pieces come together. The tree says, "Wow! Now I can grow!!!" Fragrances come to me from the bark. Things shift, shimmy, and shake. I'm standing there! Everything happens at once! The Life Energy just takes off! It's like fireworks! Once that shift occurs, it's only a matter of time before the result is visible in the tree.　　　Dr. Jim Conroy, The Tree Whisperer

The Five Heralds
BE the Tree
ASK the Tree
HEAL the Tree
(SAVE the Tree)
LOVE the Tree

TRY THIS **AGAIN: SAVE THE TREE** OR **PLANT**

Read through the steps. Decide whether you feel comfortable doing this exercise. If you do not feel comfortable or have any concern whatsoever, do not do it.

Note: Saving the tree or plant begins after you have given it the Healing Whisper. A shift of its bioenergy—from pulling in to pushing out—may occur. Healing drives growth. Growth drives healing. Be patient. Give the tree a chance to heal.

Return to a tree often that received the Healing Whisper.

At all times: Be patient. Maintain a caring connection with the tree or plant. Inspect it regularly and carefully. Take pictures.
Watch for the tiniest signs of healing or new growth. Things may happen such as leaking may start to dry up. Tiny, new leaves may appear.

Step 1: Check your notes on its appearance when you did the Healing Whisper. Write down notes about its current appearance and progress below.

Step 2: Put your hands on the Green Being and BE THE TREE. Use one of the imagining or breathing techniques from earlier in this notebook/journal.

Step 3: ASK THE TREE, **"Are you continuing to heal?"** _____

_____

_____

Step 4: ASK THE TREE, **"Do you want the Healing Whisper again?"**
❑ *If "yes," ask when, and do it now, or whenever it says to do it. See pages 64, 66, or 68.*

Step 5: ASK THE TREE, **"Is there something else you need?"**
❑ *If "yes," go through the list in the ASK THE TREE section of this chapter. Emphasize items such as adding organic compost. Make notes about what it needs below.*
❑ *If "no," ask if it wants to be left alone while it continues to heal.*

Step 6: ASK THE TREE for some intuitive insights about what is going on for it.
Be patient. The intuitive insight may take some time to come to you. Write down your intuitive impressions below.
Note: This step will give you some clarity as to what additional questions you might ask, or about doing the Healing Whisper, or if you should leave it alone while it continues to heal.

Step 7: ASK THE TREE if it needs some attention other than what you are capable of providing—including professional attention. ASK THE TREE if you should contact Dr. Jim at *www.TheTreeWhisperer.com* or a qualified local arborist.

Step 8: Thank the tree for letting you into its world. You may also write your comments, reflections, impressions, and notes here.

_____

_____

_____

_____

_____

_____*continues...*

Healing of internal functionality drives growth. Growth drives healing of internal functionality. Fully operational, healthy trees and plants are actively growing. When a sick tree or plant is HEALED and actively growing again, it has been saved from decline and death. To complete SAVING THE TREE, you must be patient and calm. You must remain positive and connected to it. It can take more time than you think depending on how sick the tree was initially. You may have to return to do the Healing Whisper often. ASK THE TREE how often it needs the Whisper. Please, give the tree a chance. It may take time until new, green growth appears.

DR. JIM CONROY, THE TREE WHISPERER

The Five Heralds
BE the Tree
ASK the Tree
HEAL the Tree
SAVE the Tree
(LOVE the Tree)

Try This: LOVE THE TREE—A Breathing Connection Cycle

Read through the steps. Decide whether you feel comfortable doing this exercise. If you do not feel comfortable or have any concern whatsoever, do not do it.

Audios are available for download at www.TreeWhispering.com.

Note: LOVE THE TREE occurs all the time! You may especially feel the loving experience while you are doing the initial Healing Whisper. You may also feel it when you return to the tree or plant.

If you are still with the tree or plant after giving it the Healing Whisper, begin at Step 3. If you have returned to it later, begin at Step 1.

Step 1: Put your hands on a tree or plant. Remove shoes and/or close eyes, if comfortable doing so.

Step 2: Ask the tree or plant for permission to engage in this exercise with you.

Say: **"Do I have permission to interact with you?"**

❑ *If you feel disquiet or agitation, or received a "no," please stop. You do not have permission.*

❑ *If you feel calm or peaceful, or received a "yes," continue here.*

Step 3: Imagine a stream of air coming into your heart area from the tree and going out your belly area back to the tree. This stream of air creates a connection cycle.

❑ *On the incoming stream of air, imagine the Life Force and bioenergy of the tree connecting with you.*

❑ *On the outgoing stream of air, imagine your Life Force and bioenergy connecting with the tree.*

Step 4: Continue to make the cycle of connection. Allow the tree's Life Force to come in at your heart area and allow your Life Force to go out to the tree from your belly area. Breathe in and out easily and gently. Allow yourself to enjoy this connection cycle for several minutes.

Step 5: Feel the purity of your shared connection with the tree or plant. Feel the back and forth flow of love between you and the tree or plant.

Step 6: Acknowledge that your connection is loving the tree or plant.

Step 7: Thank the tree or plant for allowing you into its world. Write down any comments, impressions, feelings, or notes.

_____

_____

_____

_____

_____

_____*continues...*

The pure connection I feel with a tree during the healing process is best described as a profoundly humbling feeling of LOVE.

My appreciation of the complex and beautiful ways trees and plants operate has expanded exponentially since I started my tree-healing work.

My gratitude for the Spirit that trees contain goes beyond words.

DR. JIM CONROY, THE TREE WHISPERER

The Five Heralds
BE the Tree
ASK the Tree
HEAL the Tree
SAVE the Tree
(LOVE the Tree)

TRY THIS AGAIN:

LOVE THE TREE—A BREATHING CONNECTION CYCLE

Read through the steps. Decide whether you feel comfortable doing this exercise. If you do not feel comfortable or have any concern whatsoever, do not do it.

Note: LOVE THE TREE occurs all the time! You may especially feel the loving experience while you are doing the initial Healing Whisper. You may also feel it when you return to the tree or plant.

If you are still with the tree or plant after giving it the Healing Whisper, begin at Step 3. If you have returned to it later, begin at Step 1.

Step 1: Put your hands on a tree or plant. Remove shoes and/or close eyes, if comfortable doing so.

Step 2: Ask the tree or plant for permission to engage in this exercise with you.

Say: "**Do I have permission to interact with you?**"

❏ *If you feel disquiet or agitation, or received a "no," please stop. You do not have permission.*

❏ *If you feel calm or peaceful, or received a "yes," continue here.*

Step 3: Imagine a stream of air coming into your heart area from the tree and going out your belly area back to the tree. This stream of air creates a connection cycle.

❏ *On the incoming stream of air, imagine the Life Force and bioenergy of the tree connecting with you.*

❏ *On the outgoing stream of air, imagine your Life Force and bioenergy connecting with the tree.*

Step 4: Continue to make the cycle of connection. Allow the tree's Life Force to come in at your heart area and allow your Life Force to go out to the tree from your belly area. Breathe in and out easily and gently. Allow yourself to enjoy this connection cycle for several minutes.

Step 5: Feel the purity of your shared connection with the tree or plant.
Feel the back and forth flow of love between you and the tree or plant.

Step 6: Acknowledge that your connection is loving the tree or plant.

Step 7: Thank the tree or plant for allowing you into its world. Write down any comments, impressions, feelings, or notes.

---

continues...

⟨⟩ *I really do love plants and I am grateful to be with them so much.*
     My key memory from the Tree Whispering class is from the first evening. Basia led us in a guided visualization where we touched a plant and imagined feeling what it would be like to be inside of its leaves and roots.
Then, Dr. Jim led us outside to put our hands on the majestic White Pines at the entryway. I've been a gardener for many years, but I never before felt this. Suddenly, I could feel my hands go inside of the tree, like I've heard about psychic healers doing with people. My hands were inside of this tree, touching the Life Force of this tree. Now, I visualize this happening and practice it more and more. Tree Whispering has opened my eyes and my senses to what is possible to perceive.
     ADAELA MCLAUGHLIN, HAVERHILL, MASSACHUSETTS, PROFESSIONAL GARDENER AND LANDSCAPER

⟨⟩ *I always ask for confirmation of whether or not the tree or plant received the healing. I sometimes get confirmation in miraculous ways. One day in autumn, I was communicating with and healing my favorite Beech tree. My car was parked not far away from the tree and the car door was open. Later, driving away, I noticed that a Beech nut had landed down at the bottom of the console between the two front seats. At that point, I say "thank you."*
     DR. JIM CONROY, THE TREE WHISPERER

Because most of my plants are herbs, when I plant, I want to do the best I can for each plant right from the beginning. My first step is to make a connection with the Life Force and Growth Energy of the plant while it is still potted.

Using the Transplant Chore technique, I ask it for information about the best location, orientation, and depth in the ground. I find that some don't care, that they just want to be in the soil. Others are more particular, asking for a certain position in the garden and certain distance from other plants. Still others want to be turned a certain way and set-in at a certain depth.

Considering the intelligence of Nature within each plant makes me pay attention to the fact that all species are not alike. I pause, ask, and appreciate each one.

LINDA FARMER, MASSACHUSETTS, HERBALIST, GARDENER, AND
LONG-TIME TREE WHISPERING STUDENT

Could a plant tell you "no"? It could.

And if it does, I would say to you, "respect that answer." Hearing a "no" answer does not have to hurt your self-confidence or your agenda!

The plant's "no" answer may be a guidance system leading you to a different way to accomplish what you are trying to do. A "no" answer may indicate that the plant needs you to ask different questions. Use your professional knowledge to ask more specific questions.

Ask the plant's bioenergy field and innate Intelligence some more questions. It could mean "not now, but maybe later," or it could mean "do it another way." For examples: "Don't transplant me here, put me there" or "Organic compost in two weeks, not now."

Generally, the plants are willing to work with us. They are not like stubborn children. Getting a "no" doesn't have to mean "absolutely no!" You will probably be able to do what you want, but it may not be your way. Be open to a different way. Let the plant or tree explain it to you with this yes-or-no language or by intuitive knowing.

Trust the path.

My client discovered a leak in her basement in August. She had to move a medium-sized tree that was growing where the repair needed to be made. August just happens to be the worst time to move a tree because of the extreme heat.

I connected with the tree and explained the situation to the tree in advance. I asked its permission. I got a "no."

I re-explained the situation with more detail. I said the owner would move the tree where the tree wanted to be moved. I said that I would brief the owner on exactly how the tree wanted to be cared for after the transplant.

Once again, I asked permission. I got a "yes." The transplanting was good to go.

I treated the tree prior to transplanting to bring it to optimal health. Then, I explained what would happen to it and when. I kept my promise and did what I said I would do about placement and follow-up. The tree thrived even with a hot August transplant.

And, here's a last note on being told "no." I have found that many trees that have been under stress need to recuperate before they can tolerate transplanting or a product. It's like us: If you have just had the flu, you might not want to sit down to a steak and potatoes dinner right away.

Trust the path.

DR. JIM CONROY, THE TREE WHISPERER

Part 5: A Gift for You—
The Holistic Chore™ for Transplanting

Dr. Jim starts: Basia and I offer a workshop called the Holistic Chores™. These are holistic, bioenergy-based approaches to the practical activities you have to do anyway, like transplanting, removing unwanted plants, applying products, harvesting, or preparing for construction.

As a gift, we offer you the Holistic Chore for Transplanting.

Of course, you need to use commonly accepted good practices for transplanting your particular tree or plant and situation. We give you this gift to use in combination with good practices.

Basia teaches: In the "Stepping Inside Their World" exercise, you probably felt what it might be like to stand in one place, have roots in the soil, have leaves waving in the wind. That's called "coming from the tree's or plant's point of view." The Holistic Chores' holistic, bioenergy approach asks you to expand that level of awareness up to empathy or deep emotional understanding of the problems and challenges that the Green Being is encountering in its life and in its environment.

The main idea behind the Holistic Chores is respect. You would not like it if someone uprooted you—so it is only respectful to give advance notice of your intention to the Green Being and ask it for permission.

The beauty of the Holistic Chores is that you form a partnership with the Green Being. Then, you use methods to alert it to your intentions and help it prepare itself. Your communications and actions can have a real impact in helping a plant reprioritize or redirect its bioenergy flow and avoid going into shock.

Dr. Jim informs: There are two parts to the Holistic Chore for Transplanting—and that only makes sense. In the first part, you meet with the tree or plant before the transplanting—a few hours to a week before, depending on the age of the plant. You explain the situation, ask it for permission, then ask it other questions about being transplanted. Offering the Healing Whisper™ will help to prepare it. In the second part, you partner with the plant just before and right after the transplanting to minimize its shock.

Basia reminds about boundaries: I have to say this again for your protection. When you connect with a tree's or plant's bioenergy, you receive information about its health or get messages from it. You should be receiving positive information and pleasant sensations, not pain or discomfort of any kind. If you feel pain or distress, stop the process immediately. Develop strong, healthy personal and psychological boundaries before you communicate with a tree or plant, or work to heal a sick one.

HOLISTIC CHORES™
Practical Activities Done in Partnership and Cooperation with Trees and Plants

PART 1-ADVANCE NOTICE OF TRANSPLANTING:

Read through the steps. Decide whether you feel comfortable doing this exercise. If you do not feel comfortable or have any concern whatsoever, do not do it.

Audios are available for download at www.TreeWhispering.com.

Learn the particular good practices for transplanting the tree or plant you intend to move. Larger trees or plants should be transplanted professionally, but watch that they use good practices. You will do more harm to the tree or plant by transplanting it incorrectly. Poor transplanting leads to stress and decline. Maintain good practices for your particular tree or plant and its new location.

Review: Consider returning to the "Stepping Inside Their World" guided meditation exercise to make the connection with the tree or plant.

Step 1: Go to the tree or plant that you intend to transplant. Sit comfortably or stand with the tree or plant. All steps may take 10 minutes or more.

Step 2: Ask permission to connect/interact in the following way:

❑ *Breathe slowly and deeply. Close your eyes, if possible.*

❑ *Imagine your own heart's biofield. Imagine the tree's or plant's bioenergy field.*

❑ *Ask Permission:* "**May I connect with you?**"

If you feel peaceful or get a "yes," then you have permission.

If you get a "no" or feel disquiet or agitation, then you don't have permission. Drink some water. Check on your own level of attentiveness. Then, try asking again.

❑ *If you get a "no" again, honor it. Stop your interaction. Don't force yourself in.*

❑ *Upon getting a "yes," establish your bioenergy connection with the Green Being or experience the two biofields overlapping. Allow yourself to enjoy, see, hear, smell, sense, feel, or intuit. Relax into "coming from the Green Being's point of view." You may receive information or thoughts. You may feel sensations.*

Step 3: Give advance notice.

❑ *For larger trees or plants, you will be giving them 1 to 7 days notice. But ask!*

❑ *For small plants or houseplants, a few hours to a day is good. But ask!*

❑ *Explain—with compassion—why you want to transplant the tree or plant. Speak either from your heart or aloud.*

❑ *Explain the steps that will happen to the tree or plant from your heart.*

In your heart or aloud, say to the tree or plant:

"I want to transplant you on [day/date] _____ because [give your reason] _____

_____."

"What will happen to you is [brief description] _____

_____

_____."

continues...

Step 4: Ask the tree or plant: **"Do I have your permission to transplant you?"**

❏ *If you feel disquiet or you get a "no" in your intuition, find out why.*

❏ *Do your best to set aside your human judgment and open yourself to come from the plant's or tree's point of view. Ask it for more insight into why it is telling you "no." Ask more questions. Look for alternatives. Trust the path.*

❏ *Reasons for a "no" might include: Another time of the season might be better for it to acclimate; it may have healthier roots on one side that it doesn't want cut; it may be under stress from recent conditions and would not survive the transplant now. ASK!*

❏ *If you feel a positive feeling or hear/see a "yes," continue.*

Step 5: Say to the tree or plant:
"I plan to return to transplant you on [day/date]_____. Is that enough time for you to prepare yourself for the move?"

❏ *If you get a "no," ask incrementally about longer amounts of time: "one day?", "two days?", "another week?", "another month?", etc.*

❏ *If you feel a positive feeling or get a "yes," continue here.*

NOTE: If you have little to no flexibility about the timing, say instead:
**"I plan to return to transplant you on [day/date]_____.
Can you prepare yourself by that time?"**

Step 6: Ask permission and give the tree or plant the Healing Whisper:
Ask: **"Do I have your permission to give you a Healing Whisper?"**

❏ *If "no," skip to Step 7.*

❏ *If "yes," continue here:*

• *Hold the Green Being and confirm your connection with its Life Force.*
• *Touch the Green Being with one hand.*
• *With the other hand, touch or tap on the roots and stem.*
• *Say 4 to 8 times aloud or in your heart while tapping:*
"Nature Consciousness: Please remove blockages and distribute this Green Being's Growth Energy where it is needed."

❏ *Ask: "Did you fully receive the Healing Whisper?"*

❏ *If you get a "no" or have an awareness that it did not receive the treatment, repeat the tapping and the Healing Whisper.*

❏ *If you get a "yes" or have an awareness that it did receive the treatment, continue.*

Step 7: Be open to receive a message from the tree or plant you are planning to transplant. Also, be open to messages from surrounding trees or plants.

Step 8: Say **"thank you"** to the tree or plant and its Spirit from your Heart.

Step 9: You and the tree or plant have made a cooperative agreement about the time and conditions for the transplanting. Keep your word; return when you said you would.

Step 10: Jot notes about your experience or messages. _____

_____

_____*continues...*

(Part 2 of Transplanting continues on next page.)

PART 2–DURING AND AFTER TRANSPLANTING:

Read through the steps. Decide whether you feel comfortable doing this exercise. If you do not feel comfortable or have any concern, do not do it.

Note: Learn the particular good practices for transplanting the tree or plant you intend to move. Larger trees or plants should be transplanted professionally, but watch that they use good practices. You will do more harm to the tree or plant by transplanting it incorrectly. Poor transplanting leads to stress and decline. Maintain good practices for your particular tree or plant and its new location.

Review: Consider returning to the "Stepping Inside Their World" guided meditation exercise to connect with the tree or plant.

Step 1: Go to the tree or plant with which you already completed Part 1 of the Transplanting Chore. Sit comfortably or stand with the tree or plant. All steps may take 10 minutes or more.

Step 2: Ask permission to connect/interact in the following way:

❑ *Breathe slowly and deeply. Close your eyes, if possible.*

❑ *Imagine your heart's biofield. Imagine the tree's or plant's bioenergy field.*

❑ *Ask Permission:* **"May I connect with you?"**

If you feel peaceful or get a "yes," then you have permission.

If you get a "no" or feel disquiet or agitation, then you don't have permission. Drink some water. Check on your own level of attentiveness. Then, try asking again.

❑ *If you get a "no" again, honor it. Stop your interaction. Don't force yourself in.*

❑ *Upon getting a "yes," establish your bioenergy connection with the Green Being or experience the two biofields overlapping.*

Step 3: Confirm the tree's or plant's preparation.

Note: Harsh weather conditions since the original advance notice could have prevented the Green Being from fully preparing itself, and/or it might now be stressed.

Say in your heart: **"I asked permission to transplant you on [day/date]_____. You gave permission. I asked you to prepare yourself. Are you prepared?"**
❑ *If you feel disquiet or you get a "no," find out why. Ask the tree or plant for more insight into why it is telling you "no." Do your best to set aside your human judgment and open yourself to come from the plant's or tree's point of view. Act accordingly. If necessary, consider alternatives.*

_____

_____

_____

_____

_____

_____

❑ *If the Green Being answered "yes," to the question "Are you prepared?" then, continue here.*

continues...

Step 4: Say to the tree or plant:

"Please pull your bioenergy and your Life Force into your trunk/stem and main roots. After you are transplanted, I will tell you to release your energy back into your whole structure and into your new location."

Step 5: Wait until you feel an inner sense of knowing that this "pulling in of bioenergy" has happened. *You may ask, "Have you pulled in your Life Force?"*
❏ *If "no," wait and little longer. You may repeat Step 4.*
❏ *If "yes," continue here.*

Step 6: Be open to receive a message from the tree or plant that you are planning to transplant. Also, be open to messages from surrounding trees or plants.

Step 7: Do the transplanting using best practices for your tree or plant and for your location. Or supervise others, such as professionals who may be doing the transplant for you.
Remember to water frequently after transplanting according to the plant's needs.
Note: If you are accustomed to talking to your trees or plants as part of your usual care process, feel free to give them reassuring communications during the transplant process.

Step 8: Return as soon as possible after the transplant. Reconnect as described in Step 2 on the previous page.

Step 9: Ask permission to release the bioenergy of the plant.
Say in your heart or aloud: **"The transplanting process is complete. Please release your bioenergy back into your whole structure. Please release your bioenergy into your new location."**

Step 10: Wait until you feel an inner sense of knowing that this release or upwelling of bioenergy has happened.
❏ *You may ask, "Have you released your Life Force into your whole structure/new location?"*
❏ *If "no," wait and little longer. You may repeat Step 9.*
❏ *If "yes," continue here.*

Step 11: Ask permission to give the tree or plant the Healing Whisper.
Ask: **"Do I have your permission to give you a Healing Whisper?"**
❏ *If "no," skip to Step 12.*
❏ *If "yes," continue here:*

   • *Hold the Green Being and confirm your connection with its Life Force.*
   • *Touch the Green Being with one hand.*
   • *With the other hand, touch or tap on the roots and stem.*
   • *Say 4 to 8 times aloud or in your heart while tapping:*
   "Nature Consciousness: Please remove blockages and distribute this Green Being's Growth Energy where it is needed."
❏ *Ask: "Did you fully receive the Healing Whisper?"*
❏ *If you get a "no" or have an awareness that it did not receive the Healing Whisper, repeat the tapping and the Healing Whisper.*
❏ *If you get a "yes" or have an awareness that it did receive the treatment, continue:*

Step 12: Ask the tree or plant: "Is there something else you need?" and wait to receive its answer. Confirm your first thought or feeling with a "yes/no."

Step 13: Be open to perceiving a sign of confirmation such as sudden wind blowing, appearance of something, a scent from the Green Being, etc.

Step 14: Be open to receive a message from the tree or plant that you have transplanted. Also, be open to messages from surrounding trees or plants.

Step 15: In your heart, say **"thank you"** to the tree or plant and its Spirit.

Step 16: Jot a few notes about your experience or messages you may have received from the tree or plant in this notebook/journal especially if the tree or plant wants additional care.

Step 17: Water regularly or do proper after-care! Visit the plant and check on it. Repeat the Healing Whisper process from earlier in this notebook/journal, whenever needed, and do anything practical that is appropriate. Continue with good growing practices. If the tree or plant is still not doing well, ask it if it wants the Healing Whisper again, or professional help.

I have no preference or agenda when I approach a tree. More than that, I have learned to check my ego at the door. Therefore, I respect a "no." Hearing "no" does not challenge my personality, and I am not afraid to hear it. I know that trees are not like petulant children. "No" is a piece of good information that leads me to be more creative and insightful because I have to keep asking more carefully worded and deeper questions until I get a better understanding. A "yes" leads me to the next phase of questioning or leads me to clarity. Sometimes a "yes" or "no" doesn't initially fit my idea of the way I think it should be, but later on in the process, it all becomes clear. I must trust the path. Dr. Jim Conroy, The Tree Whisperer

HOLISTIC CHORES ™
Practical Activities Done in Partnership
and Cooperation with Trees and Plants

TRY THIS **AGAIN**: PERFORM THE
HOLISTIC CHORE FOR TRANSPLANTING

PART 1 AGAIN - ADVANCE NOTICE OF TRANSPLANTING:

Read through the steps. Decide whether you feel comfortable doing this exercise. If you do not feel comfortable or have any concern whatsoever, do not do it.

Audios are available for download at www.TreeWhispering.com.

Learn the particular good practices for transplanting the tree or plant you intend to move. Larger trees or plants should be transplanted professionally, but watch that they use good practices. You will do more harm to the tree or plant by transplanting it incorrectly. Poor transplanting leads to stress and decline. Maintain good practices for your particular tree or plant and its new location.

Review: Consider returning to the "Stepping Inside Their World" guided meditation exercise to make the connection with the tree or plant.

Step 1: Go to the tree or plant that you intend to transplant. Sit comfortably or stand with the tree or plant. All steps may take 10 minutes or more.

Step 2: Ask permission to connect/interact in the following way:

❑ *Breathe slowly and deeply. Close your eyes, if possible.*

❑ *Imagine your own heart's biofield. Imagine the tree's or plant's bioenergy field.*

❑ *Ask Permission:* **"May I connect with you?"**

If you feel peaceful or get a "yes," then you have permission.

If you get a "no" or feel disquiet or agitation, then you don't have permission. Drink some water. Check on your own level of attentiveness. Then, try asking again.

❑ *If you get a "no" again, honor it. Stop your interaction. Don't force yourself in.*

❑ *Upon getting a "yes," establish your bioenergy connection with the Green Being or experience the two biofields overlapping. Allow yourself to enjoy, see, hear, smell, sense, feel, or intuit. Relax into "coming from the Green Being's point of view." You may receive information or thoughts. You may feel sensations.*

Step 3: Give advance notice.

❑ *For larger trees or plants, you will be giving them 1 to 7 days notice. But ask!*

❑ *For small plants or houseplants, a few hours to a day is good. But ask!*

❑ *Explain—with compassion—why you want to transplant the tree or plant. Speak either from your heart or aloud.*

❑ *Explain the steps that will happen to the tree or plant from your heart.*

In your heart or aloud, say to the tree or plant:

"I want to transplant you on [day/date] _____**because [give your reason]** _____

_____."

"What will happen to you is [brief description] _____

_____

_____."

continues...

Step 4: Ask the tree or plant: **"Do I have your permission to transplant you?"**

❑ *If you feel disquiet or you get a "no" in your intuition, find out why.*

❑ *Do your best to set aside your human judgment and open yourself to come from the plant's or tree's point of view. Ask it for more insight into why it is telling you "no." Ask more questions. Look for alternatives. Trust the path.*

❑ *Reasons for a "no" might include: Another time of the season might be better for it to acclimate; it may have healthier roots on one side that it doesn't want cut; it may be under stress from recent conditions and would not survive the transplant now. ASK!*

❑ *If you feel a positive feeling or hear/see a "yes," continue.*

Step 5: Say to the tree or plant:
"I plan to return to transplant you on [day/date]_____. Is that enough time for you to prepare yourself for the move?"

❑ *If you get a "no," ask incrementally about longer amounts of time: "one day?", "two days?", "another week?", "another month?", etc.*

❑ *If you feel a positive feeling or get a "yes," continue here.*

NOTE: If you have little to no flexibility about the timing, say instead:
**"I plan to return to transplant you on [day/date]_____.
Can you prepare yourself by that time?"**

Step 6: Ask permission and give the tree or plant the Healing Whisper:
Ask: **"Do I have your permission to give you a Healing Whisper?"**

❑ *If "no," skip to Step 7.*

❑ *If "yes," continue here:*

• *Hold the Green Being and confirm your connection with its Life Force.*
• *Touch the Green Being with one hand.*
• *With the other hand, touch or tap on the roots and stem.*
• *Say 4 to 8 times aloud or in your heart while tapping:*
"Nature Consciousness: Please remove blockages and distribute this Green Being's Growth Energy where it is needed."

❑ *Ask: "Did you fully receive the Healing Whisper?"*

❑ *If you get a "no" or have an awareness that it did not receive the treatment, repeat the tapping and the Healing Whisper.*

❑ *If you get a "yes" or have an awareness that it did receive the treatment, continue.*

Step 7: Be open to receive a message from the tree or plant you are planning to transplant. Also, be open to messages from surrounding trees or plants.

Step 8: Say **"thank you"** to the tree or plant and its Spirit from your Heart.

Step 9: You and the tree or plant have made a cooperative agreement about the time and conditions for the transplanting. Keep your word; return when you said you would.

Step 10: Jot notes about your experience or messages. _____

_____

_____*continues...*

_____(Part 2 of Transplanting continues on next page.)

PART 2 AGAIN – DURING AND AFTER TRANSPLANTING:

Read through the steps. Decide whether you feel comfortable doing this exercise. If you do not feel comfortable or have any concern, do not do it.

Note: Learn the particular good practices for transplanting the tree or plant you intend to move. Larger trees or plants should be transplanted professionally, but watch that they use good practices. You will do more harm to the tree or plant by transplanting it incorrectly. Poor transplanting leads to stress and decline. Maintain good practices for your particular tree or plant and its new location.

Review: Consider returning to the "Stepping Inside Their World" guided meditation exercise to connect with the tree or plant.

Step 1: Go to the tree or plant with which you already completed Part 1 of the Transplanting Chore. Sit comfortably or stand with the tree or plant. All steps may take 10 minutes or more.

Step 2: Ask permission to connect/interact in the following way:

❑ *Breathe slowly and deeply. Close your eyes, if possible.*

❑ *Imagine your heart's biofield. Imagine the tree's or plant's bioenergy field.*

❑ *Ask Permission:* **"May I connect with you?"**

If you feel peaceful or get a "yes," then you have permission.

If you get a "no" or feel disquiet or agitation, then you don't have permission. Drink some water. Check on your own level of attentiveness. Then, try asking again.

❑ *If you get a "no" again, honor it. Stop your interaction. Don't force yourself in.*

❑ *Upon getting a "yes," establish your bioenergy connection with the Green Being or experience the two biofields overlapping.*

Step 3: Confirm the tree's or plant's preparation.

Note: Harsh weather conditions since the original advance notice could have prevented the Green Being from fully preparing itself, and/or it might now be stressed.

Say in your heart: **"I asked permission to transplant you on [day/date]_____.
You gave permission. I asked you to prepare yourself. Are you prepared?"**

❑ *If you feel disquiet or you get a "no," find out why. Ask the tree or plant for more insight into why it is telling you "no." Do your best to set aside your human judgment and open yourself to come from the plant's or tree's point of view. Act accordingly. If necessary, consider alternatives.*

_____

_____

_____

_____

_____

_____

❑ *If the Green Being answered "yes," to the question "Are you prepared?" then, continue here.*

continues...

Step 4: Say to the tree or plant:

"Please pull your bioenergy and your Life Force into your trunk/stem and main roots. After you are transplanted, I will tell you to release your energy back into your whole structure and into your new location."

Step 5: Wait until you feel an inner sense of knowing that this "pulling in of bioenergy" has happened. *You may ask, "Have you pulled in your Life Force?"*
❏ *If "no," wait and little longer. You may repeat Step 4.*
❏ *If "yes," continue here.*

Step 6: Be open to receive a message from the tree or plant that you are planning to transplant. Also, be open to messages from surrounding trees or plants.

Step 7: Do the transplanting using best practices for your tree or plant and for your location. Or supervise others, such as professionals who may be doing the transplant for you.
Remember to water frequently after transplanting according to the plant's needs.
Note: If you are accustomed to talking to your trees or plants as part of your usual care process, feel free to give them reassuring communications during the transplant process.

Step 8: Return as soon as possible after the transplant. Reconnect as described in Step 2 on the previous page.

Step 9: Ask permission to release the bioenergy of the plant.
Say in your heart or aloud: **"The transplanting process is complete. Please release your bioenergy back into your whole structure. Please release your bioenergy into your new location."**

Step 10: Wait until you feel an inner sense of knowing that this release or upwelling of bioenergy has happened.
❏ *You may ask, "Have you released your Life Force into your whole structure/new location?"*
❏ *If "no," wait and little longer. You may repeat Step 9.*
❏ *If "yes," continue here.*

Step 11: Ask permission to give the tree or plant the Healing Whisper.
Ask: **"Do I have your permission to give you a Healing Whisper?"**
❏ *If "no," skip to Step 12.*
❏ *If "yes," continue here:*
  • *Hold the Green Being and confirm your connection with its Life Force.*
  • *Touch the Green Being with one hand.*
  • *With the other hand, touch or tap on the roots and stem.*
  • *Say 4 to 8 times aloud or in your heart while tapping:*
"Nature Consciousness: Please remove blockages and distribute this Green Being's Growth Energy where it is needed."
❏ *Ask: "Did you fully receive the Healing Whisper?"*
❏ *If you get a "no" or have an awareness that it did not receive the Healing Whisper, repeat the tapping and the Healing Whisper.*
❏ *If you get a "yes" or have an awareness that it did receive the treatment, continue:*

Step 12: Ask the tree or plant: "Is there something else you need?" and wait to receive its answer. Confirm your first thought or feeling with a "yes/no."

Step 13: Be open to perceiving a sign of confirmation such as sudden wind blowing, appearance of something, a scent from the Green Being, etc.

Step 14: Be open to receive a message from the tree or plant that you have transplanted. Also, be open to messages from surrounding trees or plants.

Step 15: In your heart, say **"thank you"** to the tree or plant and its Spirit.

Step 16: Jot a few notes about your experience or messages you may have received from the tree or plant in this notebook/journal especially if the tree or plant wants additional care.

Step 17: Water regularly or do proper after-care! Visit the plant and check on it. Repeat the Healing Whisper process from earlier in this notebook/journal, whenever needed, and do anything practical that is appropriate. Continue with good growing practices. If the tree or plant is still not doing well, ask it if it wants the Healing Whisper again, or professional help.

A "yes" or "no" is not a definitive answer. It is a direction. It gives you a place to go forward from. It helps guide you along the path so that you are truely coming from the plant's point of view and getting the answer that is appropriate for the plant. When I am in the flow of getting answers to my question, the "yes" or "no responses also give a deep insight about how important it is to ask questions. DR. JIM CONROY, THE TREE WHISPERER

The best part for me is giving voice to a Being who can't speak for itself. We exist in a world with many other Beings, we all need to work together, to listen to each other, all of us.
CHERYL SMITH, PHD, UNIVERSITY OF NEW HAMPSHIRE, EXTENSION PROFESSOR AND PLANT HEALTH SPECIALIST

In my youth, I got an arborist's degree but never worked in the field. After retiring, I took the Tree Whispering workshop and immediately felt empowered. I know that the workshop changed my life. Dr. Jim and Basia suggested to all the students that we get involved with the trees in our towns. I did! I brushed up my arborist's license, then I got the town to create a tree commission. It wasn't long before I got myself appointed as the tree commissioner.

As the tree commissioner, one of my main goals is to have people look at and touch the trees. I think people can feel close to trees while they are doing a tree inventory in the town. Volunteers gather data about the size of our town's tree canopy and its health. Knowing such data can help the town save money. But, the most important thing is that the people get out, touch the trees, and get to know them.

So, I am positioning myself for the time when I can bring up what I learned even more. I want to save trees through this awareness that Dr. Jim and Basia teach. It's my mission.

People who used to joke with me are starting to change their thought processes. People are calling me for my advice because the word is out that I will give sincere advice about the care of their trees. They know that one of my tools is feeling the energy of the trees. They have seen me take out my little "cheat sheet" from class.

People are looking for other options. They love the trees and they are starting to accept my actions more. They enjoy hearing that this tree energy is—in fact—there. The majority have told me that they believe it.   LEO G. KELLY, WEST HAVEN, CONNECTICUT, ARBORIST AND MASTER GARDENER

My husband and I spent four years challenging a misguided school in my town who proposes to cut down a thousand trees in one of the remaining historic, old-growth forests in New Jersey. Building sports fields is not more important than preserving a forest. Instead, the forest could be a prized resource for environmental studies.
I am passionate about trees. I feel that as living Beings they have a right to stay in their place and grow. The kind of future that I want to see is where people wake up to the benefits that trees provide us and give the trees equal protection.
SALLY MALANGA, WEST ORANGE, NEW JERSEY, BUSINESS OWNER AND TREE PROTECTOR

By going to the Tree Whispering workshop, I experienced trees with the other students as a group. By being together, I felt a stronger current. It was freeing for me to see that each person was finding her or his own way of connecting and no way was wrong. One might have shoes on, another had shoes off, one might be touching the tree while another was touching the ground, one might be hugging a tree, while another stood away and left some space. At the same time, being in the group gave me a sense of belonging and way to feel the sensations of connection more strongly because all the attention from all the students was focused more strongly.
MADELINE "GROWEESHA" THOMPSON, BOONTON, NEW JERSEY, BUSINESS OWNER AND PROFESSIONAL COUNSELOR

What can you do to feel good about your connection with Nature when you are around other people? A big step in confidence is simply starting a conversation with your friends about trees. Ask them whether they have a favorite tree or if they remember a special tree from their childhood. You will start people talking about how much they love trees—just like you. You'll be amazed! You will find many others like yourself. It's fun!   BASIA ALEXANDER, THE CHIEF LISTENER

COURAGE. Noun. The quality of mind or Spirit that enables a person to face difficulty, danger, pain, etc., without fear.

Synonyms: audacity, bravery, backbone, daring, determination, dauntless, fearlessness, fortitude, gallantry, gameness, guts, heart, heroism, intrepidity, lion-heartedness, mettle, nerve, pluck, Spirit, stoutheartedness, temerity, tenacity, undaunted, valiance, valor.
Roget's II: The New Thesaurus and Dictionary.com

Part 6: Belonging and Ambassadorship

Dr. Jim suggests: By this point, you probably have notes from your experiences and messages from the trees or plants written in your *Tree Whispering: Trust the Path Notebook and Journal*. Now, you can be their spokesperson. You are their ambassador. Who else do they have to spread their stories and their wisdom?

If you get a message or experience from one or more trees or plants, it is a gift. Like any gift, you may want to go out and share it. That way, you become an ambassador for Green Beings. By being their ambassador, you are one with all. We are all connected. Bless you for your courageous work.

Basia adds encouragingly: Whether you simply share a personal experience with a friend, get involved with your town—like Leo did—or go all the way into the kind of courageous activism that Sally spearheaded, I believe that sharing will feel good to you. It's not that scary, really, once you do it. Other people will appreciate your candor and valor. Each of us who steps out and talks about love of trees, the experience of healing trees' inner health, and trees' wise messages, will contribute to the shift in consciousness growing on our dear planet now.

Go to *www.TreeWhispering.com* and let us know what you've done. Otherwise, email or call us. Contact information is at the back of this book. You might receive an award or your words could be quoted in the next book.

TRY THIS: FEEL CONNECTED WITH OTHER NATURE LOVERS

Read through the steps. Decide whether you feel comfortable doing this exercise. If you do not feel comfortable or have any concern whatsoever, do not do it.

Step 1: Sit down in a comfortable and private place where you won't be interrupted for five to ten minutes. This exercise is best done with eyes closed, if you are comfortable closing your eyes.

Step 2: Millions and millions of people around the world are just like you in their enjoyment, caring, respect, warmheartedness, and love for trees, plants, and all of Nature. Take some time to feel or think about having something in common with a tremendous group of fellow Human Beings.

Step 3: Imagine feeling a sense of belonging. We are all connected.

Step 4: Jot down a few notes about your experience. Short phrases are okay.

TRY THIS: IMAGINE BELONGING

Read through the steps. Decide whether you feel comfortable doing this exercise. If you do not feel comfortable or have any concern whatsoever, do not do it.

Audios are available for download at www.TreeWhispering.com.

Step 1: Sit down in a comfortable, private place without interruption for 10 to 15 minutes. This exercise is best done with eyes closed.

Step 2: Groups go by many names. Use this list or find your own words that represent "belonging," "acceptance," or "compatibility" to you.

❑ *Family*	❑ *Friends*	❑ *Association*	❑ *Community*
❑ *Relations*	❑ *Companions*	❑ *Club*	❑ *Neighborhood*
❑ *Tribe or Clan*	❑ *Fellowship*	❑ *Team or Crew*	❑ *Congregation*
❑ *"My People"*	❑ *Kindred Spirits*	❑ *Co-workers*	❑ *Convention*
❑ *Cousins*	❑ *Sister/Brother*	❑ *Partners*	❑ _____

Step 3: For a minute, allow yourself to feel an emotion of self-appreciation and self-acceptance as a member of such a group.

Step 4: When you are ready, change the scene. Imagine that you are standing or sitting in a town plaza, a field, a forest, or a hotel ballroom. There is a bright light or the sun is shining on you. It is very comfortable, pleasant, and easy for you.

Step 5: Imagine that there are other people there. These people are similar to you in their respect and caring for trees, plants, and Nature.
You like them. They like you. All of you appreciate and honor Nature.

Step 6: Imagine enjoying yourself with these people. Everyone is smiling, making greetings, and honoring everyone else. It's a very safe place.

Step 7: Imagine that more people, who are similar to you, arrive.

Step 8: Imagine being comfortable, happy, and excited that hundreds, thousands (or even millions) of people who are just like you, are in this place. All of you are celebrating your appreciation for the Plant Kingdom and kinship with trees, plants, and all of Nature.

Step 9: Imagine that your favorite tree, other trees and plants, as well as other Beings of Nature come to this place.

Step 10: Imagine that the trees, plants, and other Beings of Nature are rejoicing together with you. It is a very safe place.

Step 11: Take a few moments to deeply enjoy and find pleasure in your feelings of acceptance and belonging with these groups of people and Beings of Nature.

Step 12: Change the scene any way you want to, if that will give you more pleasure and more sense of belonging. Enjoy yourself.

Step 13: When you are finished, write notes and reflections about your experience.

_____

_____

_____

_____*continues...*

Having a larger perspective can help. Millions of people around the world are similar to you in their enjoyment, caring, respect, or love for trees. They, like you, have a special sensitivity to the Life Force of trees and plants. Consider that you have something in common with many other people.   BASIA ALEXANDER, THE CHIEF LISTENER

TRY THIS: MUSTER COURAGE TO BECOME AN AMBASSADOR FOR TREES AND PLANTS

Read through the steps. Decide whether you feel comfortable doing this exercise. If you do not feel comfortable or have any concern whatsoever, do not do it.

This exercise may involve a particular tree or plant with which you already did a "Try This" exercise earlier in the notebook/journal.

Step 1: Think about the range of people who you know.

- ❏ *Family*
- ❏ *Friends*
- ❏ *Neighbors*
- ❏ *Children*
- ❏ *Other parents*
- ❏ *Clubs you belong to*

- ❏ *Business associates or coworkers*
- ❏ *People at your place of worship*
- ❏ *People who do services for you*

- ❏ *Garden center manager*
- ❏ *Town officials*
- ❏ *Alumni or current classmates*
- ❏ *Newspaper editor*
- ❏ *Online groups*

❏ *[Add your own]*_____

Step 2: Make two lists of people using the categories below. Write their names.

People with whom I can share trees' and plants' messages as the message truely came to me:	*People for whom I must adjust the wording of the message so that it seems more neutral:*
_____	_____
_____	_____
_____	_____
_____	_____
_____	_____
_____	_____
_____	_____

Step 3: Think about the kind of courage it will take to share messages.

COURAGE: Noun. The quality of mind or Spirit that enables a person to face difficulty, danger, pain, etc., without fear. Synonyms: audacity, bravery, backbone, daring, determination, dauntless, fearlessness, fortitude, gallantry, gameness, guts, heart, heroism, intrepidity, lion-heartedness, mettle, nerve, pluck, Spirit, stoutheartedness, temerity, tenacity, undaunted, valiance, valor

❏ Will you have to find the right moment?

❏ Will you have to explain something first, such as mentioning that you are doing an exercise from a book you are reading?

Step 4: Designate yourself the trees' or plants' ambassador.

Say to yourself, "**I am a Tree and Plant Ambassador.**"

How does it feel to be an Ambassador? _____

_____

_____

_____*continues...*

Step 5: What messages did a specific tree or plant want you to share on its behalf?

Note: You may want to go back and read through the messages you have received or you may want to go to a specific tree and ask it what message it would like you to share on its behalf.

_____

_____

_____

Step 6: Do the sharing!

Here are some suggestions for sharing the tree's or plant's message.

❏ *Talk directly about your personal experience with someone of your choosing. Begin by sharing something from your personal point of view, then share the communication you received from the specific tree or plant.*

❏ *Talk about the messages the tree or plant wanted you to share on its behalf.*

❏ *Talk about the needs or health requirements of the tree or plant to someone who can do something about them.*

❏ *Write a note, article, or letter to someone. You may also post it online.*

❏ *Post the message you got at www.TreeWhispering.com.*

❏ *Do something creative: tell a story, paint, draw, compose music, do a play, etc.*

❏ *Lead others—especially children—to do something meaningful and/or creative.*

❏ *Bring people back to the tree so they can experience something themselves.*

❏ *[Add your own]_____.*

Step 7: Jot a few notes about what you thought or how you felt while you shared.

_____

_____

_____

Step 8: You may go back to the Green Being. Tell it from your heart what happened when you shared its messages. You are its ambassador. Ask the Green Being how you did. Then, ask it for suggestions about how you might improve at being its ambassador. You might be surprised at the answer.

_____

_____

Step 9: Go to *www.TreeWhispering.com* and let us know what you've done. You might receive an award or your words could be quoted in the next book. Email us at *Messages@TreeWhispering.com.* Also see other contact information at the back of this book.

🌱 *I contacted our town's new alderwoman about the trees. She asked me to write an article. I said 'yes' even though I didn't know how to write it. Then, I thought I would ask the trees what they would like to say—through me. I have ideas but ultimately, I am their ambassador.*
MARY CYPRESS, RACINE, WISCONSIN, STREET ARTIST

Try This Again: Muster Courage to Become an Ambassador for Trees and Plants

Read through the steps. Decide whether you feel comfortable doing this exercise. If you do not feel comfortable or have any concern whatsoever, do not do it.

This exercise may involve a particular tree or plant with which you already did a "Try This" exercise earlier in the notebook/journal.

Step 1: Think about the range of people who you know.

❑ Family
❑ Friends
❑ Neighbors
❑ Children
❑ Other parents
❑ Clubs you belong to

❑ Business associates or coworkers
❑ People at your place of worship
❑ People who do services for you

❑ Garden center manager
❑ Town officials
❑ Alumni or current classmates
❑ Newspaper editor
❑ Online groups

❑ [Add your own]_____

Step 2: Make two lists of people using the categories below. Write their names.

People with whom I can share trees' and plants' messages as the message truely came to me:

People for whom I must adjust the wording of the message so that it seems more neutral:

_____   _____

_____   _____

_____   _____

_____   _____

_____   _____

_____   _____

_____   _____

Step 3: Think about the kind of courage it will take to share messages.

COURAGE: Noun. The quality of mind or Spirit that enables a person to face difficulty, danger, pain, etc., without fear. Synonyms: audacity, bravery, backbone, daring, determination, dauntless, fearlessness, fortitude, gallantry, gameness, guts, heart, heroism, intrepidity, lion-heartedness, mettle, nerve, pluck, Spirit, stoutheartedness, temerity, tenacity, undaunted, valiance, valor

❑ Will you have to find the right moment?

❑ Will you have to explain something first, such as mentioning that you are doing an exercise from a book you are reading?

Step 4: Designate yourself the trees' or plants' ambassador.
Say to yourself, "**I am a Tree and Plant Ambassador.**"
How does it feel to be an Ambassador? _____

_____

_____

_____*continues...*

Step 5: What messages did a specific tree or plant want you to share on its behalf?
Note: You may want to go back and read through the messages you have received or you may want to go to a specific tree and ask it what message it would like you to share on its behalf.

Step 6: Do the sharing!
Here are some suggestions for sharing the tree's or plant's message.

❏ *Talk directly about your personal experience with someone of your choosing. Begin by sharing something from your personal point of view, then share the communication you received from the specific tree or plant.*

❏ *Talk about the messages the tree or plant wanted you to share on its behalf.*

❏ *Talk about the needs or health requirements of the tree or plant to someone who can do something about them.*

❏ *Write a note, article, or letter to someone. You may also post it online.*

❏ *Post the message you got at www.TreeWhispering.com.*

❏ *Do something creative: tell a story, paint, draw, compose music, do a play, etc.*

❏ *Lead others—especially children—to do something meaningful and/or creative.*

❏ *Bring people back to the tree so they can experience something themselves.*

❏ *[Add your own]_____.*

Step 7: Jot a few notes about what you thought or how you felt while you shared.

Step 8: You may go back to the Green Being. Tell it from your heart what happened when you shared its messages. You are its ambassador. Ask the Green Being how you did. Then, ask it for suggestions about how you might improve at being its ambassador. You might be surprised at the answer.

Step 9: Go to *www.TreeWhispering.com* and let us know what you've done. You might receive an award or your words could be quoted in the next book. Email us at *Messages@TreeWhispering.com.* Also see other contact information at the back of this book.

When I went to clients' properties, I would go to give the answers, as if I had them all. Now I go to receive answers from the trees. Rather than thinking I will pry information from these majestic and mysterious tree Beings, I simply stand or kneel, holding the tree, and wait for the information to be given as a gift.

I realized that doing Tree Whispering is all about connectedness. It's about sharing the mystery of life. It isn't about the ego. It isn't about waving my hands around and healing something. It isn't about justifying my years of formal education or my investment in my profession. Doing Tree Whispering is about receptivity—allowing an insight to come and allowing something good like a healing to be given back. When I kneel with a tree, I'm receiving more than I am giving. I am there as a fellow citizen of Earth–an equal Being–sharing a unique moment of communication with another Being.
GERRY VERRILLO, GUILFORD, CONNECTICUT, ARBORIST

I've climbed and worked with trees most of my life, but that first evening in the workshop was an amazing expansion of my conscious awareness of trees. We touched Maples first. I was present to my Maple's subtle twisting movements with a deeper understanding of what was happening for the tree than ever before. I realized that I had always been looking at trees, but not seeing them. All of a sudden, the Maple's movement became visible to me. It was very peaceful and beautiful. I felt an intimate connection with the tree.

I'd always seen trees move in different ways based on the forces involved or from the way that I would move them as an arborist. But, this was very different. I feel that the way the tree was moving with a light breeze was something that was always there but I'd never seen before. I was seeing how the wind and the movement in the canopy translated into the trunk. It was surprising that there was a sort of a twisting motion. I'd never perceived that before, but it made perfect sense. It was Divine in its beauty and its trueness to its growing conditions.
DAVID SLADE, GUILFORD, CONNECTICUT, ARBORIST AND BUSINESS OWNER

Native Americans call trees "Grandmother" or "Grandfather." I was taught herbalism by a Cherokee medicine priest. Now I routinely communicate with the herbs I offer to clients to be sure that the herbs resonate with the client's body.

My Cherokee medicine priest teacher taught me to deeply honor and respect trees. He explained that he would ask a tree if there was anything he could do for it, and then he would give it a sacred offering. The tree would engage, or not. Not all trees respond to all people. If his invitation was accepted, then he explained that communication would begin.

I was learning how to meditate while I happened to be sitting next to an old, gnarled Cottonwood. The tree took me inside of it. I went into the center of the trunk. I was on a journey. My consciousness went up and down simultaneously. I heard the vibrations of the leaves. It was like hearing a million tiny violins all playing together, yet I felt that I could hear each one individually. Then, I went into the roots and heard this amazing gong sound. After that, I was inspired to continue meditating. Basia and Dr. Jim gave me a similar opportunity to journey by teaching me to step inside the plant's world in their way.
CATHY, NEW JERSEY, ACUPUNCTURIST AND HERBALIST

Developing perceptive abilities or intuition is like developing a natural resource. It exists but lays dormant until it is found. We use our ability to connect with people and things all the time, we're just not consciously aware of that. We sense things and feel things all the time; we often take that for granted. When we make efforts to connect consciously with the Life Energy of a tree or plant, we find our dormant intuitive skill.

Your intuitive sense uses imagination; it uses specific metaphors that you—and only you—are going to understand. It uses the concepts to which you already personally relate. In this way, the intuitive sense aims you toward the information you will need.

At the same time, it's important for you to have good, personal boundaries. You need to know what to allow into your body, mind, and Spirit. You need to keep out what is not good and what does not serve you. Be discerning.
MADELINE "GROWEESHA" THOMPSON, BOONTON, NEW JERSEY, BUSINESS OWNER AND PROFESSIONAL COUNSELOR

Just by touching them, I feel the energy flow in trees. I don't know if this ability has increased, but I think I have more of an awareness of it. I have always worked with the land, but now I am noticing the trees as individuals. Touching the trees fills me up; it's a real spiritual connection. When I really connect, I can focus and breathe. It fills me up mind, body, and Spirit. It's like a meditation, but I don't like to just sit. So, connecting gets me away from all the stuff in life and helps me be clear.

JUDE VILLA, MARTHA'S VINEYARD, MASSACHUSETTS, PROFESSIONAL LANDSCAPER AND DESIGNER

I walk up to a tree and state my intention quietly. I say, "I'd like to come inside. I'd like to listen to you." Then, I'll touch the tree with fingertips of both hands, then with my forehead or with my nose. Sometimes I feel electric impulses. I feel the pulses coming to me and then going back into the tree. Sometimes I don't ask questions; I just sense what I am feeling.
MIKE NADEAU, SHERMAN, CONNECTICUT, HOLISTIC LAND CARE PRACTITIONER

What do I do to perceive a tree? I simply go up to a tree to meet it. Within minutes, I can feel the tree's energy, usually in my heart or solar plexus, and sometimes I feel a pulsing in my hands. Painting, for me, has been a great way to develop a relationship or connection to Nature and to trees in particular. Learning the Tree Whispering techniques gave me a good structure for this practice. I greet the tree, ask permission, check-in with it, and then I have information.
                                                      JUDY PERRY, MAINE, ARTIST AND LANDOWNER

I used to be skeptical that a tree may or may not have anything to say, and I was judgmental of myself. I decided to stop criticizing myself. I realized that I had to be forgiving about getting messages from the trees. I have since expanded my perceptions by being totally open and nonjudgmental. Now I let myself receive whatever is conveyed to me. I do this by not expecting to get a certain thing. The best way to get better at perception is do it more often.
CARRAIG "ROCKY" ROMEO, PRINCETON, NEW JERSEY, BUSINESSMAN AND DRUID

I don't feel an energy connection through my body, but I know that I communicate with my trees and plants. I think of it more as talking to someone. Then, I pause and hear an answer. My advice is about going back to basics. Keep your eyes open. Be aware of what's happening as you go for a walk in the woods. Pay attention beyond the little circle just around you.

JOAN LENART, NORTH HAVEN, CONNECTICUT, GARDENER AND HOMEOWNER

Try participating in practices that expand awareness of body, such as Native American rituals or yoga. It might feel a little strange at first, but awkwardness is okay. Just get through it. Don't be embarrassed. Welcome the gift of the moment. Share that part of yourself with other people. See a bigger picture.

CHERYL SMITH, PHD, UNIVERSITY OF NEW HAMPSHIRE, EXTENSION PROFESSOR AND PLANT HEALTH SPECIALIST

Sometimes we prevent ourselves from hearing correctly; we hear what we want to hear, not what we need to hear. I used to resist, but now I accept perceptions as they come. I simply recognize the fear and get past it. Letting go of cultural training and my own resistance is important for self growth and for connecting with all the other Spirits out there—both Human Beings and Nature Beings. When I let go, the limiting things I believe go away. I allow what I feel to come to me and accept that it is okay, that it is right.

DANIELLE ROSE, NORTHERN NEW JERSEY, AUTHOR, NEWSPAPER COLUMNIST

Trees and plants may not be "talking" to me, but they are expressing their essence of Being. I appreciate that trees are living, breathing organisms that are crucial to our world. In my work, I have learned how to help plants thrive. After the workshop, my relationship with all plants has deepened and I have found that trees and plants can teach me, too. Now, a stronger connection with trees will truly allow the healing process to take place.

DORRIE ROSEN, RLA, NEW YORK, NEW YORK, PLANT INFORMATION SPECIALIST, NEW YORK BOTANICAL GARDEN

Since the class, I don't think that my perceptive abilities have changed, but I think that my willingness to have them change has increased. That's a big step for me. When I think about myself before the class, I was taking trees for granted. I saw them as objects, even though I knew they were living Beings. One of the big gifts is that I don't cut trees down anymore. My thinking is "who am I to decide to cut down this living tree?"

DWIGHT BROOKS, KATONAH, NEW YORK, ARBORIST, HORTICULTURALIST, ORGANIC LAND CARE TEACHER

☼ ☼ *I love to lean against my favorite tree—a Gingko in the West Village area of New York*
☜☉☞ *City. I go into the connection with the Gingko without any expectations. I respect it by*
asking permission to communicate. I encounter the Being knowing—within it—there is a
unique consciousness and Life Force. While energetically connected, my physical sensations have
to feel good to me. Then, I like to be playful! I let the ideas come, and I know whether they are right
or wrong by whether the energy-feeling I have is stronger or weaker.
MARIA PETROVA, NEW YORK, NEW YORK, PROFESSIONAL GRAPHIC DESIGNER AND ENERGY MEDICINE HEALER

🔯   *My advice is to develop conscious conversation with the Kingdoms of Nature. If we allow
     ourselves to activate the connection between the Plant Kingdom and our bodies, it is much
easier to communicate. Don't try to figure it out. It just takes patience and attention. You're
learning a language that is already being spoken inside of you.*
                    ILONA ANNE HRESS, L.C.S.W., C.M.T., REV., MADISON, NEW JERSEY, SPIRITUAL HEALER*

As a Maple syrup producer, I sense the trees' needs. I suppose you could say that we have a partnership; we are equals. I think that I have a knack for nurturing plants that comes from a deep appreciation of them. What is really interesting is the more I know about trees, the less I know about them. I just get to a place where I believe I understand, and then they sneak in a total surprise that throws off my knowledge.

CHUCK WINSHIP, EAST SPRINGWATER, NEW YORK, MAPLE SYRUP FARMER

As a citizen of our green environment, I aspire to be a comforter and healer for these gentle giants.

ALEXANDRA SOTERIOU, PASSAIC COUNTY, NEW JERSEY, ENERGY HEALER AND BUSINESS OWNER

All people are immersed in a sea of information that comes to us through the nonlinear mind. And, as humans, we have the unique capability to draw upon that subtle knowledge and to add to it our logic, our book-learning, and our intentions.
We become empowered to go—with conviction—into action. We get information, then our very next question has to be, "What do I do with this information?" Moving into action is the joining of intuition with logic, knowledge, and even insights from others for the betterment of all. In taking action steps, we use all of our abilities; we can make any combination of conventional and nonlinear, heart-oriented steps.   MARY MCNERNEY, LINCOLN, MASSACHUSETTS, ATTORNEY

Afterword: Your Trail Guides

Dr. Jim Conroy,
The Tree Whisperer®

Ms. Basia Alexander,
The Chief Listener

WHO ARE THESE PEOPLE?

Dr. Jim Conroy, The Tree Whisperer®, says: I'm not going to give you much of a resume. I don't think you want a list of successes or accomplishments; I think that you want to know who I am—on the inside—what I believe and what I stand for.

First and foremost, I love trees and plants. Trees and plants are alive like you and me. We are all living Beings. We are just different. By listening to them, I have come to know hundreds and hundreds of trees and plants as individuals. When they are sick, I help them heal themselves. I love each of them as if they were my children.

I received a doctorate degree from Purdue University in Plant Pathology then went into the corporate world for over 25 years. But, I made a 180 degree turn from those days in the chemical industry when I realized that conventional ways are focused—not on the inner health of these living Beings—but on putting things on the outside while coming from a very human-centric point of view. When I did my years in the industry, no one ever bothered to ask the trees, plants, insects, disease organisms, or other living Beings of Nature *themselves* how to help them become healthier. In those days, professionals thought they knew best and gave no credit to the wisdom of Nature.

Now, when I "whisper" with trees and plants, I am getting to know a tree or plant as an individual. I am coming from its point of view. Mostly, I am asking it about what conditions must be corrected in its internal functionality in order to improve its health. Yes, trees and plants contain that information since they are living systems, but they can't always accomplish the change in internal functionality for themselves. It's just like us—we might know about our own illnesses, but we can't always heal ourselves.

Communicating with the Beings of Nature and healing them when they are sick is my passion and my life's mission. I feel a calling to educate, too. I want to inform people that their relationship with trees, plants, and other Beings of Nature can be a two-way street. But, it's more than just information. I don't like to lecture. Instead, Basia and I invite everyone into their own personal experience of connection with the Life Force of Green Beings. Doing is the best way to learn.

Anyone and everyone can open their hearts and minds to make contact with the sacredness in all living Beings. All people can—and should—connect with these Beings in order to expand appreciation, extend gratitude, and co-create harmony, peace, and balance in Nature.

Ms. Basia Alexander, The Chief Listener, says: I am a "big picture" person. I like to have the perspective on where humanity has been and where it is going. And, where are we headed? Well, I am also an optimist so I am convinced that—despite our problems—humanity is headed for the best times yet.

I believe that we have the opportunity to individually rise above our sense of limitation and leap into a higher consciousness. I believe we can mutually cooperate and co-create good times and a healthy planet with our partners in Nature and with each other.

I love to teach. I love to learn. The two go together: I cannot teach without learning from my students! It is a dream-come-true for me to be able to invite people to learn about and experience listening to the wisdom and intelligence in Nature. To me, that is the highest form of practical spirituality.

Tree Whispering is far more than a business partnership for Dr. Jim and me. We know that people the world over love and depend on their trees and plants. We hold the vision of active and meaningful collaboration between people and the Plant Kingdom for the good of all.

About the Authors

Dr. Jim Conroy, The Tree Whisperer®
Whisperer@TheTreeWhisperer.com

Ms. Basia Alexander, The Chief Listener
Listener@TreeWhispering.com

P.O. Box 90, Morris Plains, NJ 07950 USA

*Dr. Conroy and Ms. Alexander offer classes and workshops in Tree Whispering®
and Cooperative BioBalance®. Through these educational experiences, people
gain paradigm-shifting principles and easy, simple practices for restoring inner
health to trees and plants, while feeling good in body, mind, and Spirit.*

TreeWhispering.com and CooperativeBioBalance.org.

*<u>Dr. Jim Conroy, The Tree Whisperer</u>®, earned his PhD in Plant Pathology from
Purdue University, Indiana. After 25 years as a plant and tree health senior
executive in top 50 agricultural chemical companies, he did a 180° turn-around.
He started to come from the plant's point of view. Now, he is an authority on
Nature-based communication and a global expert who holistically heals stressed
trees, plants, crops, forests, and ecosystems.*

*His breakthrough to develop integrated ecosystem healing and balancing, called
Cooperative BioBalance™, came through his research and profound spiritual
connection with the Plant Kingdom. In 2002, Dr. Jim developed the Green
Centrics™ System, a holistic, bioenergy healing, hands-on, no-product, green-
friendly, and sustainable solution to restoring tree and plant health from the
inside-out. He since created and developed Co-Existence Technologies™.*

*As a practitioner, he works on estates, golf courses, forests, botanical gardens,
municipalities, and people's properties to heal stressed or declining trees by
restoring functionality to internal parts and systems.*

*He is also a NOFA certified organic land-care specialist, educator, faculty
member at the Omega Institute and The Nature Lyceum, keynote speaker,
healer, rose grower, founder and president of Plant Health Alternatives, LLC,
and co-founder of the Institute for Cooperative BioBalance.*

TheTreeWhisperer.com, CooperativeBioBalance.org, and StrengthenForests.com.

*<u>Ms. Basia Alexander, The Chief Listener</u>, is a catalyst for positive change,
innovator, and leader in the new field of Conscious Co-Creativity. As an expert
Nature communicator, Basia develops leading-edge, synergistic concepts and
produces transformative and inspirational curriculum. She has also produced
manuscripts on topics including spiritual expansion, personal organizing,
health, and creativity. She founded ReVitalizations™—a health and creativity
coaching and training business and co-founded Trees for Tomorrow—a local
environmental action group in Wayne, New Jersey. As Training Director for an
Apple® Computer dealership and as an Adjunct Professor at Essex College, New
Jersey, Basia wrote and delivered trainings in desktop publishing. She created a
Track II Bachelor of Arts in Communications at Beloit College, Wisconsin.*

*After gaining certification as a BodyTalk System practitioner, her love of plants
led her to form a partnership in Plant Health Alternatives, LLC, with Dr. Jim.
Basia writes all Tree Whispering manuals and teaches, side-by-side, with him.
As co-founder of the Institute for Cooperative BioBalance, Basia directs its
educational and philanthropic initiatives.*

TreeWhispering.com, CooperativeBioBalance.org, and PartnerWithNature.org.

Participate

BE DESIGNATED AS A DEPUTY

Now that you have read this book, you can become a Deputy in the Peace Pipe Global Network Pattern. You will be invited to participate as Dr. Jim directs local and global tree-healing sessions. You can offer healing to the Plant Kingdom in coherence with other Deputies around the world. The sum of the work will be greater than the parts. You'll also receive updated techniques and information so you can help your own trees and plants. Go to www.TreeWhispering.com to find out how to qualify so you can get your Deputy Badge.

SEND YOUR STORY

As a result of reading this book, if you have an interesting experience or get a message from a tree or plant, please submit your story to Messages@TreeWhispering.com. Dr. Jim and Basia would like to hear it and could consider it for inclusion in future books.

TAKE A CLASS OR WORKSHOP

Enroll in Tree Whispering® classes for a deeper experience of connection with trees, plants, and other Beings of Nature. Learn the Holisic Chores™–practical activities done in partnership with trees and plants. Receive all of the Healing Whispers™–experiential bioenergy healing techniques done in cooperation with trees and plants.

To go a step further, the full Tree Whispering Workshop provides both a profound experience with Green Beings as well as a system of healing techniques. Advanced classes in Green Centrics™ and trainings for people to lead introductions are planned. Gain insight into your own health in the class Trees Are Your Healers.

Discover an idea whose time has come in the new field of study called Cooperative BioBalance®—Partnering with Nature and Balancing EcoSystems. It is the peace treaty among people, all plants, insects, diseases, and related organisms. Practice new guardianship by touching, asking, healing, saving, and loving the Beings of Nature.

FORWARD AN INITIATIVE

The Institute for Cooperative BioBalance offers initiatives in which people may participate. Contact Participate@CooperativeBioBalance.org.

INNOVATIONS, PRODUCTS, AND SPECIAL OFFERS

Check TreeWhispering.com for relevant innovations, supportive products, and special offers as they become available. These include audio downloads of the "Try This" exercises, the companion book to this notebook/journal: Tree Whispering: A Nature Lover's Guide to Touching, Healing, and Communicating with Trees, Plants, and All of Nature, *and other supplementary materials and opportunities.*

Ordering Information

Published by: Plant Kingdom Communications
P.O. Box 90, Morris Plains, New Jersey, 07950
Barbara@PlantKingdomCommunications.com

Ebook version also available.

Available wherever books are sold and through these locations:

www.PlantKingdomCommunications.com/Tree_Whispering_Book

Distributed in the US by: Atlas Books, Ashland, Ohio, 800-266-5564
http://www.atlasbooks.com/atlasbooks/ Search on "Tree Whispering"

Also available through Baker & Taylor, Ingram, New Leaf, and at your local or independent bookstore.

Price: $11.95   ISBN: 978-0-9834114-1-3
Format: Soft cover, spiral bound, 120 pages, 6 x 9 inches.

Go to www.PlantKingdomCommunications.com:
- for additional information
- to order a signed/personalized copy
- for other educational products
- to schedule talks or classes
- to set-up book signings
- to watch videos
- and more...

Other Books and Products from Plant Kingdom Communications

The Main Book:
Tree Whispering: A Nature Lover's Guide to Touching, Healing, and Communicating with Trees, Plants, and All of Nature
ISBN: 978-0-9834114-0-6

MP3s of the "Try This" Exercises:
Available at www.TreeWhispering.com

The Tree Whispering Kit:
Includes the main book, this notebook/journal, and a wearable item.
ISBN: 978-0-9834114-7-5

Find information about Tree Whispering classes and workshops at
www.TreeWhispering.com.

Highlights of the authors' work to heal trees, plants, crops, forests, and entire ecosystems can be found at www.TheTreeWhisperer.com,
www.CooperativeBioBalance.org,
www.StrengthenForests.com, and www.PartnerWithNature.org